Development of Solar Power Generation and Energy Harvesting

About the Centre

The Centre for Science and Technology of the Non-Aligned and Other Developing Countries (NAM S&T Centre) is an inter-governmental organisation with a membership of 48 countries spread over Asia, Africa, Middle East and Latin America. Besides this, 11 S&T agencies and academic/research institutions of Bolivia, Brazil, India, Nigeria and Turkey are the members of the S&T-Industry Network of the Centre. The Centre was set up in 1989 to promote South-South cooperation through mutually beneficial partnerships among scientists and technologists and scientific organisations in developing countries. It implements a variety of programmes including international workshops, meetings, roundtables, training courses and collaborative projects and brings out scientific publications, including a quarterly Newsletter. It is also implementing 7 Fellowship schemes, namely, NAM S&T Centre Research Fellowship, Joint NAM S&T Centre – ICCBS Karachi Fellowship, Joint CSIR/CFTRI (Diamond Jubilee) - NAM S&T Centre Fellowship, Joint NAM S&T Centre – ZMT Bremen Fellowship, Research Training Fellowship for Developing Country Scientists (RTF-DCS), NAM S&T Centre – U2ACN2 Research Associateship in Nanosciences and Nanotechnology and Joint NAM S&T Centre – DST (South Africa) Training Fellowship on Minerals Processing and Beneficiation in Indian institutions. These activities provide, among others, the opportunity for scientist-to-scientist contact and interaction, training and expert assistance, familiarising the scientific community on the latest developments and techniques in the subject areas, and identification of technologies for transfer between member countries. The Centre has so far brought out 75 publications and has organised 107 international workshops and training programmes.

For further details, please visit www.namstct.org or write to the Director General, NAM S&T Centre, Core 6A, 2nd Floor, India Habitat Centre, Lodhi Road, New Delhi-110003, India (Phone: +91-11-24645134/24644974; Fax: +91-11-24644973; E-mail: namstcentre@gmail.com; namstct@bol.net.in).

Development of Solar Power Generation and Energy Harvesting

— Editors —

Engr. Muhammed Musa Gaji

Dr. Abhishek Verma

CENTRE FOR SCIENCE & TECHNOLOGY OF THE NON-ALIGNED AND OTHER DEVELOPING COUNTRIES (NAM S&T CENTRE)

2018

DAYA PUBLISHING HOUSE®

A Division of

ASTRAL INTERNATIONAL PVT. LTD.

New Delhi – 110 002

ISBN: 9789389569049 (Int. Edition)

Centre for Science and Technology of the Non-Aligned and Other Developing Countries (NAM S&T Centre)
Core-6A, 2nd Floor, India Habitat Centre, Lodhi Road,
New Delhi-110 003 (India)
Phone: +91-11-24644974, 24645134, Fax: +91-11-24644973
E-mail: namstct@gmail.com
Website: www.namstct.org

Published by : **Daya Publishing House®**
A Division of
Astral International Pvt. Ltd.
– ISO 9001:2015 Certified Company –
4736/23, Ansari Road, Darya Ganj
New Delhi-110 002
Ph. 011-43549197, 23278134
E-mail: info@astralint.com
Website: www.astralint.com

Digitally Printed at : **Replika Press Pvt. Ltd.**

Foreword

Ever increasing population combined with continuous technological advancement has led to a great demand for energy which continues to grow. Current, conventional non-renewable energy sources based on fossil fuels are being depleted. There is now growing global concern over impending energy crises, which, combined with man-made climate change, are driving countries to develop and adopt alternative and clean energy technologies. Renewable energy sources and technologies can provide eloquent solutions to the problems being faced, in particular in developing countries. The relative merits of renewable energy vary greatly depending on the scale, capacity, and status of individual technologies, natural resource availability, characteristics, location and a number of other factors. It is generally true that renewable energy resources are infinitely available in nearly every region of the world. The main issue is the conversion efficiencies for harnessing them and the costs involved in the processes.

Of all the energy sources, solar power holds the greatest potential for supplying all our energy needs for the foreseeable future. Solar energy has emerged as a popular choice and the desire to switch to this clean energy technology has driven significant advancements in this area. The advent of solar technology offers a clean, renewable and sustainable energy for the future. However, the requirement for cost reductions in the conversion of solar energy into electrical energy, development of storage and handling in the local environment and other related issues remain as challenges.

I am very happy that the Centre for Science and Technology of the Non-Aligned and Other Developing Countries (NAM S&T Centre) has realized how important it is to deliberate on these issues and organised an excellent International Workshop on *'Trends in Solar Power Generation and Energy Harvesting"* jointly with the Amity University, UP, India -Dubai Campus in Dubai during March 2017.

The present publication of the NAM S&T Centre comprises 20 papers from 17 developing countries and reflects the status of solar energy and the government efforts in their countries. I am sure this book will serve as an important document for all those associated with solar energy research, development and its use.

Prof. Trystan Watson

College of Engineering
Swansea University, UK

Preface

Solar Energy is playing a pivotal role in compensating the need of electrical energy, due to escalation in demand and decline trends of conventional source of energies, like coal, petroleum, natural gases. To efficiently meet out the future energy demands and energy securities; the alternative energy sources should be aggressively investigated. Therefore, an effective energy solution should be able to address long-term issues by utilizing alternative and renewable energy sources. Renewable energy resources can definitely improve the quality of life of mankind by promoting sustainable development and systems. Solar power especially, is the best choice for sustainability and renewable energy in developing countries, as it reaches more competitive levels with other renewable energy sources, may serve to sustain the lives and families of millions of underprivileged peoples in developing countries. It is more practical, reliable, cost-effective, and healthier for people and the environment.

The recent trends in solar energy generation and harvesting are to decrease the cost of the energy generation, either by introducing the low cost processing techniques or to increase the efficiency of the solar cells. After 1st and 2nd generation of bulk silicon based solar cells and thin-film Si/CdTe/CIGS based solar cells, respectively, the 3rd generation technologies are underway. Many new technologies include photo-electrochemical cells, polymer solar cells, quantum dot, tandem/multi-junction solar cells, up-conversion and down-conversion, surface plasmonic, nano-crystal solar cells and other novel innovations and inventions.

Presently, there is an urgent need for efficient harnessing of various forms of solar energy, *i.e.*, Solar Thermal and Photovoltaic in huge capacities with efficient technologies demonstrated for lightning needs, potable water generation and distribution, mid-day meal preparation and distribution in schools, vaccine preservation, health care, information distribution and comfort needs of aging population in NAM (Non-Aligned Movement) and other developing countries.

In order to deliberate on the current trends and scenario in solar power generation, its storage, harnessing and related issues for academia and industry, the Centre for Science and Technology of the Non-Aligned and Other Developing Countries (NAM S&T Centre) is bringing out this issue.

The aim was to define an approach for Solar Power Generation and intersecting themes for enabling better informed policy-making. The purpose and objective of this meeting were also to spread the latest scenario of solar energy generation and harvesting in every possible field for academia and industry.

The Book "Development of Solar Power Generation and Energy Harvesting" comprises of scientific contributions, latest scenario in NAM and developing countries and various policies. The contribution has been made by different researchers, eminent scientists and top ranked dignitaries from all over the NAM and developing countries, which includes Afghanistan, Cambodia, Cuba, Egypt, Gambia, India, Indonesia, Iran, Iraq, Malaysia, Mauritius, Morocco, Nepal, Nigeria, Palestine, Sri Lanka, South Africa, Tanzania, Togo, Turkey, United Kingdom, Zambia and Zimbabwe. The chapters include various latest and significant topics, *i.e.*, status and trends of solar power generation in various NAM countries, solar thermal systems and their industrial applications, fabrication and development in thin-film solar cells, simulations and quantitative analysis of solar cells and modules, shading effects on MPPT of PV systems, PV in Indian scenario, recent developments in organic solar cells, hydroelectric cells, DC and AC microgrids using hydrogen storage systems, Perovskite solar cells, Self-cleaning of solar panels, systems to turn buildings into powerstations, optimization for solar cell design in some chalcogenide compounds, X-Ray Switching In Commercial Grade Solar Panel, Organic- Inorganic Quantum Dots Hybrid Nanostructures for Solar Energy Harvesting, CdTe nanostructure for solar cells, Nanomaterials enhanced PCM composites for Solar thermal energy storage applications, Dye-sensitized solar cells, Lithium-ion batteries for efficient energy storage, thermal cooling layer - fabrication, characterization and testing for photovoltaic applications, designing of solar cells, risk assessment of solar power plants, and much more.

Apart from discussed fields, this book can also help in understanding various other points,such as, how the governments in few developing countries can undertook appropriate policy measures to promote renewable energy development including solar PV and Solar thermal technologies and applications, appropriate for local conditions of economic development, social development and resources availability; how strong enabling policy framework can be evolved to promote an environment where renewable energy technologies do not face unfair disadvantages compared to conventional energy technologies to nurture the local markets; how some governmental incentives strengthen the solar PV market that may positively affected installations for developing nations, *etc.*

The editors wish to place on record our appreciation to NAM S&T Centre for providing an opportunity to edit and contribute in publishing, such a book.

Engr. Muhammed Musa Gaji

Energy Commission of Nigeria,
Department of Renewable Energy,
Abuja, Nigeria

Dr. Abhishek Verma

Assistant Professor III,
Amity Institute of AdvancedResearch and Studies (Materials and Devices),
Amity Institute of Renewable and Alternate Energy,
Amity University, Noida (UP), India

Introduction

The global demand for energy is currently growing beyond the limits of installable generation capacity. To efficiently meet the future energy demands, energy security and reliability needs to be improved and alternative energy sources required to be more aggressively investigated. An effective energy solution should be able to address long-term issues by utilising alternative and renewable energy sources. Of the many available renewable sources of energy, solar energy is clearly a promising option as it is abundantly available at most places and is also the cleanest energy resource on our planet. Solar power, especially as it reaches more competitive levels with other energy sources in terms of cost, may serve to sustain the lives of millions of underprivileged people in developing countries.

The recent trends are to decrease the cost of the energy generation either by introducing the low cost processing techniques or by enhancing the efficiency of the solar cells. After the 1st and 2nd generation of the bulk silicon based - and thin-film Si/CdTe/CIGS based solar cells, the 3rd generation technologies are underway. The new technologies include photo-electrochemical cells, polymer solar, quantum dot, tandem/multi-junction, up-conversion and down-conversion, surface plasmonic and nano-crystal solar cells and other novel innovations and inventions.

In order to deliberate on the current trends in solar power generation, its harnessing, storage, and related issues, the Centre for Science and Technology of the Non-Aligned and Other Developing Countries (NAM S&T Centre) jointly with the Amity University, UP, India - Dubai Campus organised an International Workshop on **'Trends in Solar Power Generation and Energy Harvesting'** in Dubai during 27-29 March 2017.

The Workshop was attended by 33 senior experts and professionals from 23 countries. There were speakers from 20 NAM member countries including Afghanistan, Cambodia, Cuba, Egypt, The Gambia, India, Indonesia, Iran, Iraq, Malaysia, Mauritius, Morocco, Nepal, Nigeria, Palestine, Sri Lanka, South Africa,

Tanzania, Togo, UAE, Zambia and Zimbabwe as well as from two other countries, namely, Turkey and the United Kingdom.

The present book is a follow up of the above workshop and comprises 20 scientific papers by the authors from 17 countries covering several issues related to the status and trends of solar power generation, designing of solar cells, risk assessment of solar power plants and many other topics. I have great appreciation for the efforts put in by the editors Engr. Muhammed Musa Gaji and Dr. Abhishek Verma for the technical editing of the manuscripts.

I am also grateful to Prof. Trystan Watson of the College of Engineering, Swansea University, UK for writing a Foreword for this publication. The involvement and valuable efforts of the entire team of the NAM S&T Centre, especially of Dr. Kavita Mehra, Mr. M. Bandopadhyay and Ms. Meenu Galyan at all the stages of the publication process and taking all the necessary actions in giving a shape to this volume and of Mr. Pankaj Buttan in cover page designing, formatting and liaising with the printers, are highly noteworthy.

I am sure that this book will be an asset to the researchers, policy makers in government departments and ministries and non government organisations engaged in renewable energy related issues in the developing countries.

Prof. Dr. Arun P. Kulshreshtha

Director General,

NAM S&T Centre

Contents

Chapter 1

Solar Photovoltaic Energy in Cambodia "An Opportunity with Obstacles"

Row Vattanak and Chea Piseth

Deputy Chief of Generation Planning Office,
Generation Department, Edc, Cambodia,
E-mail: vattanak.edc@gmail.com

Only 65 per cent of Cambodians are connected to the electricity grid by 2016, and the vast majority of those without electricity are poor families living in rural areas. Those who are connected pay extremely high electricity prices. Despite a much lower income per capita in Cambodia, electricity prices range from $0.18 in Phnom Penh to above $0.50 in rural areas. Considering these expensive prices and the amount of sunlight in Cambodia, solar PV has the potential to be a very competitive energy option.

As of 2016, installed capacity in Cambodia was approximately 1,628 MW and 320 MW imported from neighboring countries. With total installed capacity of 1948, majority of this, approximately 47 per cent, came from hydropower plants, 10 per cent from diesel/Heavy Fuel Oil (HFO), 25 per cent from coal plants contributing, 1 per cent from biomass and other 17 per cent were imported from Vietnam and Thailand.

Since 2010, Electricity demand in Cambodia is growing rapidly with average annual growth of 20 per cent. To cope with this issue Royal Government of Cambodia's core policy, set out the priorities for the development of power generation as well as expanding transmission lines, especially from renewable energy sources, encourage private sector investment focusing on technical and economic efficiency and minimize environmental and social impact set the goal by 2020, all villages in Cambodia will have access to electricity supplied from the national grid and 70 per cent at least of all households having access to grid quality electricity by the year 2030.

To promote the Solar PV, since 2009, import taxes on solar PV components, and solar water heating components were reduced from 30 per cent to 7 per cent and from 15 per cent to 0 per cent, respectively. The solar power market has been predominantly driven by the electricity needs of people who are unable to access on-grid electricity. Increased solar PV installation is also stimulated by the two programs implemented by the REF and the MME, the Solar Home Systems (SHS) Program, and the Power to the Poor (P2P) Program, which are funded by the World Bank and AFD, respectively.

Based on desk research on the Cambodian electricity market and various investment opportunities for solar PV, the preliminary found that the likeliest candidates for an investment opportunity were Solar Home Systems (SHS), mini-grids for rural areas, and industrial systems for factory owners and SHS entrepreneurs were having a difficult time growing and scaling their businesses due to low-income customers and distributors of cheap but low quality SHS had given solar a bad reputation in the country.

Recently, RGC has approved license for the first solar farm with 10 MW installed capacity for supplying to Special Economic Zone that is expected to be completed by second half of 2017. Even thought, the tariff from solar PV higher than hydro and coal fired sources.

Keywords: *Government policy, Solar home system, Solar PV.*

1. Introduction

Cambodia lies entirely within the tropics, between latitudes 10 and 15 N, and longitudes 102 and 108 E and has achieved a stellar performance in terms of economic growth over the past decade, with an average annual GDP growth rate of 7 per cent from 2014 to 2016, and an all-time high of 13 per cent in 2005. This robust economic growth increased Cambodia's GDP per capita in Purchasing Power Parity (PPP) terms from an average US$1,797 per capita in 1993 to an all-time high of US$2,945 in 2013. Cambodia is expected to continue its rapid rate of GDP per capita growth, closing the gap with ASEAN peers such as Thailand and Vietnam through the expansion of social, economic and industrial development. Economic growth will be accompanied by an increase in energy demand across all sectors in Cambodia, but especially in the transportation, industry and services sectors (Kimura, 2014). Cambodia's Total Primary Energy Supply (TPES) in 2011 stood at 5.33 Mtoe, with oil representing the second-largest share of Cambodia's TPES at 26 per cent, while coal was the third-largest at 0.2 per cent, followed by hydro at 0.1 per cent. The country is dependent on imports of petroleum products having no crude oil production or oil refining facilities of its own. However, since 2012 Cambodia's electricity supply is reduced from depend on by oil to use the sources from hydro, coal and biomass instead of, as in 2016, installed capacity in Cambodia was approximately 1,628 MW and 320 MW imported from neighboring countries. With total installed capacity of 1948, majority of this, approximately 47 per cent, came from hydropower plants, 10 per cent from diesel/Heavy Fuel Oil (HFO), 25 per cent from coal plants contributing, 1 per cent from biomass and other 17 per cent were imported from Vietnam and Thailand, and 65 per cent of Cambodian are connected to the electricity grid (EDC, 2016). Rural electrification remains far from being achieved, and energy services are mainly delivered through fuel-based engines or generators to produce electricity that can then be stored in batteries, while biomass rather than electricity is used to power many small industrial processes. The current electricity cost is, ranging from US$0.18/kWh in Phnom Penh to US$0.50/kWh in rural areas (EAC, 2016). The supply of electricity currently is to meet basic demand, and expected to grow from 1,648 MW in 2016 to 1,681 MW in 2020 (EDC Master Plan, 2015). Although there is still considerable underinvestment in the sector, Electricité Du Cambodge (EDC) aims to provide electricity services to all villages by 2020 and to 70 per cent of all the rural households by 2030. To accelerate rural electrification, off-grid solar systems are viewed by the RGC as a potential solution in providing rural people with access to electricity. Thus, the RGC established the Rural Electrification Fund (REF) in 2004 to attract and encourage the private sector to invest in electric power infrastructure so that rural areas can have access to electricity for lighting, commercial use, handicraft production and other purposes for improving standards of living and general wellbeing. The RGC aims to promote renewable energy in its energy sector plan, targeting a 15 per cent share of Renewable Energies (REs) by 2015. RGC established Solar Home System (SHS) program in purpose to facilitate the remote rural household, which may not have access to the electricity network for a long period, to access electricity through SHS, and currently power demand increased rapidly average 20 per cent since 2010, and other hand Cambodia have a large manufacturing industry with over 600 garment factories and nineteen special economic zones that is the opportunity for investor to investment on solar PV and some of factories have installed solar PV. Recently RGC has approved license for the first solar farm with 10 MW installed capacity for supplying to Special Economic Zone that is expected to be completed by second half of 2017.

2. Solar Home System (SHS) in Cambodia

The REF was established in 2004 by the RGC to accelerate the development of rural electrification. In the period 2005–2012, the REF utilized funds provided by the World Bank under the Rural Electrification and Transmission Project (RETP) and the RGC's counterpart fund. The RETP was a US$46 million World Bank–funded project involving a US$40 million loan from the World Bank and US$6 million provided by an International Development Association and Global Environment Facility Grant to RGC (World Bank, 2012). The RETP aims were to (i) improve power sector efficiency and reliability, and reduce electricity supply costs; (ii) improve standards of living and foster economic growth in rural areas by expanding rural electricity supplies; and (iii) strengthen electricity institutions, the regulatory framework and the 'enabling environment' for sector commercialization and privatization. SHS is one of the sub-components of the project (roughly US$5 million allocated for this sub-component) and involved the installation of SHS in 12,000 household during the project implementation, the RGC integrated the Rural Electrification Fund (REF) into EDC to allow the Department of Rural Electrification to perform its works independently using Cambodian funding, while also continuing to receive grants and donations from external funding sources to assist in the development of rural electrification in Cambodia. In 2014 alone, EDC provided US$6 million for the operation of the REF and the implementation of three rural electrification development programs consisting of: (i) the Program for Power to the Poor (P2P); (ii) the Program for Solar Home Systems (SHS); and (iii) the Program for Providing Assistance to Develop Electricity Infrastructure in Rural Areas. According to the World Bank (2012), the purpose of the SHS Program is to facilitate remote rural households that may not have access to the electricity network for long periods to access electricity through SHS. SHS was one of the sub-components of the World

Figure 1.1: Most Solar Panels in the Kingdom are Part of Solar Home Systems or SHS (*Courtsey*: SEAC).

Bank–funded REF project. However, the project was completed in 2012. In the period 2014–2015, the REF has resumed its function under the responsibility and oversight of EDC, and has sold and installed 13,240 SHS-50 Wp to rural households in remote areas (EDC, 2015). To facilitate the purchasers, ensure that the SHS installed in rural households operate well, and collect the payback amount in instalments from the purchasers, EDC has contracted BNP Power Green (Cambodia) Co., Ltd to provide transportation, installation, collection of payback in instalments, and maintenance of 4,000 systems.

3. Opportunity of Solar PV in Cambodia

Cambodia has 5.8 peak sunlight hours a day, one of the best solar resources in the entire region, probably one of the best in the entire world (Bradly, K., 2017). Average sunshine in Cambodia is 6-9 hours per day and solar radiation is estimated at 5kwh/m2 per day. This creates a huge potential for Solar Home (SHS), Solar Photovoltaic (PV) and Concentrate Solar Oower (CSP). The total technical potential of solar power is 65Gwh/year (CRCD, 2004). According to a recent report by the Mekong Strategic Partners (MSP), the costs associated with solar power are falling rapidly, so much so that Cambodia can now, according to the study, cost-effectively strengthen its energy security by accelerating investment in solar and biomass technologies. Large hydropower dams and coal-fired plants – currently Cambodia's main energy sources – provide power for between eight and 11 cents per kilowatt-hour (kWh). According to the MSP report, solar installations above one megawatt can now provide electricity profitably for as low as 12 cents/kWh, which makes it "the most competitively priced alternative renewable energy technology". And it is only going to get better. Worldwide, costs are expected to continue falling in coming years. Increasing economies of scale, more efficient designs, and constant advances in materials and manufacturing guarantee of the price of solar generated electricity continues on a downward trend. A country with one of the best solar resources in the region – and possibly the world – and a technology that is quickly becoming more affordable: put two and two together, and it doesn't take a genius to realize that solar's potential in Cambodia is vast. Arguably, it is not being met. To date, the bulk of solar panels installed in the Kingdom belong to small Solar Home Systems (SHS) built on households in the countryside. Only a handful of larger solar energy ventures have been undertaken or are planned, all of them having been announced within the last two years.The new Coca-Cola plant in the Phnom Penh Special Economic Zone, for example, boasts a photovoltaic (PV) system in the rooftop that will eventually supply a third of the factory's electrical energy needs.

Figure 1.2: The Solar Farm of the New Coca-Cola Plant in the Phnom Penh Special Economic Zone Supplies Roughly of Third of its Electrical Consumption (*Courtesy*: Coca-Cola)/

Last month, Kamworks announced it will install 10,000 solar panels in Khmer Beverages' beer factory, an ambitious project that will feed the plant with approximately four megawatts of power. The same company recently concluded a PV power system on the rooftop of BKK1's Silvertown Metropolitan which has been hailed as the first commercial solar power system on a high-end residential tower in Cambodia.

Recently, EDC has sign power purchase agreement from 10MW Solar Farm on BOO Project in Bavet City, Svay Rieng province will be constructed in the nearest future. This project is the first and biggest Solar Farm to invest in Cambodia (EDC, 2016). Below is average Sun Radiation at provinces in Cambodia that have been studied by RichGrid Company.

Figure 1.3: The Memorandum of Understanding Signing Ceremony for the New Kamworks-Khmer Beverages Solar Venture, which is to Install 10,000 Panels in the Beer Factory Located just Outside the Capital, took Place on January 23 (Bradley, K., 2017).

Table 1.1: Average Sun Radiation at Province in Cambodia

Month	***Prey Veng***		***Svay Rieng***		***Ratanakiri***		***Mondulkiri***	
	Average		***Average***		***Average***		***Average***	
November	5.34	5.59	4.88	5.36	5.21	5.94	6.1	6.05
December	5.82		5.36		5.98		7.00	
January	6.22		6.08		6.94		7.37	
February	6.16		6.00		7.16		6.76	
March	5.54		5.84		6.47		5.94	
April	5.33		5.17		5.47		5.08	
May	4.69		4.2		4.37		4.08	
June	4.48	4.16	3.85	3.63	3.89	3.63	3.26	3.30
July	4.12		3.68		3.67		3.00	
August	3.88		3.3		3.1		2.55	
September	4.09		3.53		3.39		3.11	
October	4.23		3.77		4.12		4.58	
Average	**4.99**		4.64		4.98		4.90	

With a handful of notable solar power installations in the last two years, the sector seems to be picking up speed quickly. However, the number of large solar energy projects in Phnom Penh and the rest of the country is still surprisingly small.

4. Obstacle of Solar PV in Cambodia

The main hurdles keeping investors away is the lack of clarity in the regulatory environment. "As with every sector, the government has its role in establishing the ground rules and how it works, and the ground rules have been established, and if the government created more ground rules it will make it easier for private investment to come in (Bradley, K., 2017). To attract more investors and reduce risk in solar PV there is a need to refine investment cost, acquire sun radiation data, and mitigate social and environmental impacts to make these inventory projects more technical and economics sound and sustainable. To promote the decentralised, demand-driven approach in electrification and to facilitate private sector involvement in solar PV development, a number of serious barriers to overcome are summarised below:

High Project Costs

Solar PV is usually located in remote areas with limited access and far away from load centers, which make solar PV more expensive and less attractive than large hydropower and other alternatives.

Lack of Policy and Legal Framework

The policy and legal framework needs to be put in place, *e.g.* concessionary duties and taxes concerning imports of solar PV equipment.

Lack of Data

There is still little documented information available on the characteristics of energy markets, including their scope, potential and consumer characteristics. Few systematic studies exist for assessing the potential for solar PV resources in the country. There is also a need to conduct more detailed studies on investment into solar PV.

Institutional Capacity for Planning, Implementation and Operation

Technical knowledge and operational skills are in short supply. Lack of experience in operation and management and limited training possibilities are some of the factors causing institutional barriers. Lack of coordination among concerned stakeholders (governmental agencies, development partners, NGOs, private sector and financial institutions) also acts as another barrier in the absence of a comprehensive policy on solar PV development.

Incentive Regime for Solar PV Development

The RGC should put in place an incentive regime to give an impetus to the solar PV sector while addressing the macro level issues. Such a regime would have two components: a direct incentive to reduce the front-end costs and increase affordability; and an indirect incentive through a technical assistance initiative for pilot projects leading to awareness creation and capacity building for supporting a market infrastructure and particularly for accelerating the implementation process of the RGC and other donor programs.

Limited Access to Financing

A shift of global proportions is underway. The MPR report states that utilities, consumers and investors in developed and developing countries are switching to renewable energy to constrain environmental impacts but also because it is increasingly economic in its own right. "Hydro and coal are last century technologies, developed economies are moving towards more sustainable alternatives because they are more cost-effective (Bradley, K., 2017).

Factor in the energy goals of the Cambodian government – which include achieving electricity self-sufficiency by the year 2024 – and the falling costs of installing solar energy systems, and it becomes clear the Kingdom's business model for solar at the multi-megawatt scale is now particularly strong. However, banks have so far failed to seize opportunities in the sector. "Banks right now don't know about solar and are not interested in providing loans (Monteiller, C., 2017). Although there is considerable interest within international finance circles in "bankable" renewable energy projects in Cambodia, the lack of a clear policy and regulatory framework is inhibiting players like the United Nation's Green Climate Fund and the Asian Infrastructure Investment Bank from taking chances in the Kingdom's solar market, according to the MPR report.

Now is more positive about the financing options available and says that banks are now beginning to warm up to the idea of investing in the sector, noting that ANZ Bank actually has a very large regional fund dedicated to solar electricity projects (Bradley, K., 2017). Inadequate taxation may also be a part of the equation. Solar panels and other components of PV power systems are still subject to VAT. Given that some sustainable biomass energy products – such as sustainable charcoal briquettes – have recently been exempted from paying VAT, many consider this tax shouldn't be levied on solar energy products. Moreover, since electricity purchases from the EDC are VAT-exempt, solar energy companies cannot pass VAT costs to the end-consumer.

"VAT has a big impact on project prices. Eliminating the tax will provide significant room to make projects more profitable," says Bradley. According to him, equipment for certain projects can currently qualify for VAT exemption, but the process is very bureaucratic and slow. Sustainable energy equipment generally does not attract import duties, however, a seven per cent duty still applies for solar energy equipment. To make matters worse, certain components needed to build the systems, such as batteries, are subject to a 35 per cent import duty. The first step will be to lower import duties to seven per cent for all components needed to build PV systems (Monteiller, C., 2017).

Fighting Low Quality

Acquiring your very own solar panel in Cambodia isn't hard. Go to any of the many local markets around the country and you'll have plenty of opportunity to purchase a PV panel, some retailing for as low as $5. However, if you expect your system to function beyond the initial month, you might want to look elsewhere.

Figure 1.4: The PV Panels on the Rooftop of Silvertown Metropolitan will Provide Approximately 12 per cent of the Building's Energy Needs.

A new program run by French development agency AFD and funded by the European Union, with Dutch not-for-profit organization SNV as project coordinators, aims to tackle just that. Among the program's different awareness-raising activities, the great faith in a new off-grid solar product certification standard called the "Good Solar Label". Only panel providers that meet a certain standard of quality and offer after sales service – including at least two years of warranty – are able to display the label on their products. Six companies have been certified so far: Kamworks, Lighting Engineering and Solutions, BNP Power Green, CamSolar, NRG Solutions and Pteah Baitong.

"Having a quality certification like this helps Cambodians discriminate between high and low quality products, and will help solidify the reputation of solar as a durable and reliable energy source (Monteiller, C., 2017).

The Case for Solar

With a confusing legal and regulatory framework, lack of access to finance, a less-than-ideal taxation environment and an overabundance of bad quality products that give a bad rep to solar power, the hurdles are manifold. Fortunately, so are the opportunities and the arguments for the uptake of solar.

The MPR report found that Cambodia could add 1,000 GWh (gigawatt-hour) to its generating output by constructing 700 megawatts of utility-scale solar on 1,400 hectares of land. With this additional power, Cambodia would achieve electricity self-sufficiency as early as 2017. "Cambodia could rapidly achieve energy independence through solar, requiring very little land space with negligible environmental impact," the report reads.

The benefits of creating a solar economy in the Kingdom, however, go beyond simple energy security. Solar adoption will bolster employment. He explains that although setting up a PV system requires professionals with a very specific skills set, most of the workforce needed to actually install and maintain the systems can come in the form of unskilled labour. "Transitioning into a solar economy makes perfect sense for a jobs-economics standpoint (Bradley, K., 2017).

5. Conclusions

Energy access remains a fundamental development issue for Cambodia, as electricity costs are high in both urban and rural areas. Because of prolonged underinvestment in the electricity sector, Cambodia's electrification rate as just 65 per cent in 2016. Despite the passage of the Electricity Law more than a decade ago, Cambodia's electricity sector has not developed fast enough to meet demand in either urban or rural areas. The RGC, in its rural electrification master plan, has realised the adverse consequences of high electricity costs, as well as the importance of accelerating electricity access in rural areas. Based on the master plan, 70 per cent of households will be connected to the national electricity grid by 2030. In the medium term to 2020, the master plan foresees an increase in mini-grids from small hydropower and solar PV systems, including SHS, to provide electricity access in rural areas. About 12,000 households installed SHS in the period 2005-2012 under the Rural Electrification Fund (REF) established by the RGC to accelerate the development of rural electrification. However, the REF project was completed in 2012. In the period 2014– 2015, the REF has resumed its work and sold and installed 13,240 SHS-50 Wp to rural households in remote areas. Electricity costs in rural areas charged by current electricity providers using diesel generators can be above US$ 0.50/kWh, which provides an opportunity for SHS to enter the market. Since the RGC policies, subsidy of US$100 per SHS unit is for the people living in rural areas. These results implythat promoting SHS will provide remote areas with energy access, and also enable residents in remote areas to reduce spending on electricity, thereby increasing deposable incomes and the social wellbeing of rural communities. Anywhere, Average sunshine in Cambodia is 6-9 hours per day and solar radiations is estimated at 5kwh/m2 per day and have a large manufacturing industry with over 600 garment factories and nineteen special economic zones. This creates a huge potential for solar photovoltaic (PV) to meet the demand requirement average annual growth 20 per cent, with the GDP increase around 7 per cent per year. This is the opportunities for investor interest to investment on solar power generation in Cambodia.

The findings in this study point towards the following recommendations: although, the RGC subsidy 100$ per unit but for the people living in remote area are still difficultly to find budget for purchase SHS, especially for replace battery. The high installed system price of SHS is one of the obstacles in promoting the uptake of solar PV. It is recommended that the involved authorities such as the Electricity Authority of Cambodia (EAC), and the Department of Rural Electricity of EDC might look at the whole value chain of SHS from procurement through to instalment to ensure that transition costs are minimized in order to reduce the system price. It may be necessary to make large purchases of SHS directly from manufacturers, and create an effective and transparent procurement process in RE equipment, including solar PV and SHS, Mini-grids from solar PV. The electricity authorities might consider attracting investment in mini-grids supplied by solar PV, as these would provide economies of scale compared with SHS and fuel sources. Mini-grids supplied by solar PV systems offer lower system costs than SHS. However, EDC shall guarantee to purchase the all the excess energy by synchronize to national grid and the RGC should be encourage the investor for investment on solar PV by reduce Tax and VAT, in order to compete price with fossil fuel and import energy.

6. Acknowledgements

The authors would like to thank the Centre for Science and Technology of Non-Aligned and Other Developing Countries (NAM S&T Centre), and Amity University for inviting us to participate in the International Workshop on "Trends in Solar Power Generation and Energy Harvesting. We also would like to thank Electricté Du Cambodge (EDC) and Ministry of Industry and Handicraft for their advice and assistance.

REFERENCES

1. Kimura, S. (2014), Preparation of Energy Outlook and Analysis on Energy Saving Potential in East Asia. Jakarta: Economic Research Institute for ASEAN and East Asia.
2. EDC (2016), Annual Report for the Year 2016, Electricity of Cambodia.
3. EAC (2016), Report on power sector of the kingdom of Cambodia. Phnom Penh: Electricity Authority of Cambodia.
4. EDC Master Plan (2015), Cambodia Power Development Master Plan, Study by the Chugoku Electric Power Co., Inc.
5. EDC (2015), Report on Activities of the Department of Rural Electrification Fund for the Year 2014. Electricity of Cambodia, Phnom Penh, 2015.
6. IEA and ERIA (2013), Southeast Asia Energy Outlook. Paris, France: OECD/IEA, 2013
7. World Bank (2012), Implementation completion and results report. Report No: ICR2320.
8. CRCD (2004), Sustainable energy in Cambodia. The Cambodian Research Centre for Development (CRCD) © 2004 PO Box 2515 Phnom Penh Kingdom of Cambodia http://www.camdev.org
9. Bradley, K. (2017), https://www.b2b-cambodia.com/articles/opportunities-and-barriers-in-cambodias-solar-market/
10. EDC (2016), Report from Planning and Strategic Department for the Year 2016, Electricity of Cambodia
11. Monteiller, C. (2017), https://www.b2b-cambodia.com/articles/opportunities-and-barriers-in-cambodias-solar-market/

Chapter 2

Effect of Partial Shading on the PV Module Output

E.T.El Shenawy[1*], O.N.A. Esmail[2], Adel A. Elbaset[3] and Hesham F.A. Hamed[3]

[1]National Research Centre, Solar Energy Department, Dokki, Giza, Egypt
[2]Faculty of Engineering, Al-Azhar University, Qena, Egypt
[3]Faculty of Engineering, Minia University, Minia, Egypt
E-mail: essamahame@hotmail.com

Solar power generation is growing fast in many parts of the world and driven by cost reduction, economic incentives and the needs for meeting growth in electricity demand while reducing reliance on fossil fuels. The power output of solar photovoltaic (PV) arrays is optimal only under full irradiation conditions. However, under partial shading solar irradiance and hence the efficiency of the PV array can decrease dramatically. Also shading can be caused not only by clouds, but also by buildings, trees, soiling, dust and even PV cell cracking and ageing. The present paper studies the effect of the partial shading on the performance of a thin film PV module under climatic conditions of Cairo, Egypt. This effect was measured and evaluated according to practical measurement of the I-V and P-V curves of the PV module under different operating conditions and different shadow patterns. The results showed that the partial shading change the regular shape of both I-V and P-V curves, especially in the region of maximum power. Consequently, the output power from the module decreased according to the incomplete solar radiation reaching the PV module due to shadow patterns. The measurements have been applied to different cells partially shaded as follows; half cell (bottom, middle and top of the PV module); complete cell; and two cells at the same solar radiation level.

Keywords: *PV module characteristics, I-V measurements, Solar radiation, PV module shading, PV module power loss.*

1. Introduction

Growing interest in renewable energy resources has caused the photovoltaic (PV) power market to expand rapidly, especially in the area of distributed generation [1,2]. A PV system directly converts solar energy into electric energy. The main device for the energy conversion of a PV system is a solar cell. Cells are grouped to form modules, modules to form panels and panels to form arrays. A PV system may be either an array with one panel or a set of panels connected in series or parallel to form large a PV system with or without a tracking sun system. Although costly, the tracking system allows a more propitious instantaneous orientation for the panels in order to achieve higher values of energy capturing during sunny days due to the diverse perpendicular positions to collect the sun irradiation [3].

A solar cell is a highly nonlinear device and this characteristic is most apparent when PV modules composing an array are operating at different conditions [4]. The performance of the PV system depends on the operating conditions especially on solar irradiation, temperature, configuration and shading. The partial shading on PV array, for instances, due to a passing cloud or neighboring buildings causes not only energy loss in the conversion, but also further non-linearity on the I-V characteristics [5]. A partial shaded cell of a non-uniform illuminated PV system can be submitted to a negative voltage. If there is no protection, cell breakdowns can happen during non-uniform illumination. Hence, normally in order to protect the cells an extra pn-junction is implemented as a bypass diode. For instance, one bypass diode connected in parallel with each set of cells or with a panel is common practice as a compromise between protection and increase on the cost due to the extra pn-junction [6]. Also partial shading pattern with hasty change is not easy for the tracking of the maximum power point (MPP), because with non-uniform illumination usually there will be multiple local MPPs and they will change as fast as does the illumination [7].

In a typical series-connected PV module, the energy conversion efficiency is adversely affected if all the cells are not equally illuminated (partial shading conditions). All the cells connected in series are forced to carry the lower current produced by the shaded cells. The shaded cells will be reverse biased, acting as loads. Therefore, bypass diodes are introduced into the PV module to protect the shaded cells from localized overheating (hot-spot problem). In a commercial series-connected 72-cell PV module, there are three bypass diodes in parallel with one third of the solar cell array each. Generally, if one solar cell is shaded, one third of the power output from the PV module will be lost [8].

Since shading is one of the most common causes of a lower actual energy yield of a PV system in operation than the predicted one in the design phase, the shading analysis is a must during the PV system design procedure. Nowadays, we can find in the scientific literature models for shading losses calculation based on solving the whole current–voltage curve of the generator [9-12]. These models require as inputs all the details concerning the basic components of the generator (solar cells and bypass diodes), as well as the details about the electrical interconnections between them. Knowledge of the amount of shading on each solar cell is also required. Once the behavior of the basic components is known, the behavior of the whole generator is calculated by applying the laws of the circuit theory. This implies dealing with non-linear equations systems that must be numerically solved. The main advantage of these models is their accuracy. However, their practical application presents difficulties. First, a great deal of information is required as input in order to be able to reproduce the I–V curves of the solar cells and bypass diodes. This information is not easy to get. On the other hand, dealing with the whole I–V curve of the basic components is problematic from a computational point of view, taking into account that a minimum of 50 points is required to determine an I–V curve [13] and that a PV generator is composed of thousands of solar cells. In addition, solving the non-linear equations that govern the behavior of the generator is time consuming. Finally, the implementation of these models and the numerical algorithms is not easy. So these models are oriented to research works and not to their common use by the PV professionals in yield estimation [14].

When it is not possible to have a set of I–V curves that characterize the components of the generator, the user is forced to apply a simplified model. Existing software packages rely on simplified expressions that allow a fast estimation of the shading losses [15,16]. However, an experimental campaign carried out showed that these expressions lead to appreciable errors [17]. These authors proposed more accurate models based on an empirical equation that improves the previous methods. It shows good results in the calculation of the energy yield. However, it still gives important deviations in the calculation of the instantaneous power losses. Depending on the considered generator and its particular configuration, the authors reported Root Mean Square Errors (RMSEs) between the modeled and the experimental power losses of up to 26 per cent [18].

In this work, the performances of a thin film PV module is investigated practically under conditions where half, one or two cells in the module are partially shaded according to the climatic condition of Cairo, Egypt (Lat. 30 2′ 38″ North, Long. 31 14′ 9″ East). The performance of the PV module is evaluated by measuring a different I-V and P-V characteristics of the PV module under different operating solar radiation levels with different shading patterns applying on the surface of the PV module.

2. PV Module

The PV module (Figure 2.1) is a thin film silicon type with rated power output of 22 W at Standard Test conditions (STC, 1000 W/m^2 of solar radiation and surface temperature of 25°C). Table 2.1 shows the electrical characteristics of the PV module. The PV module is supported up on a tilted structure from steel frames. The tilt angle is fixed at 30° with horizontal and the structure is mounted such that the module is facing south direction.

3. Measuring Circuit and Shadow Patterns

To investigate the performance loss of the PV module under partial shadow, a series of partial shadow patterns are applied to the PV module at different operating solar radiation levels. The partial shadow patterns are arranged as follows;

Figure 2.1: PV Module.

Table 2.1: Electrical Characteristics of the PV Module

Item	***Specifications***
Type	Thin film
Rated power	22 W
Open circuit voltage	22 V
Short circuit current	1.8 A
Operating voltage	15.6 V
Operating current	1.4 A

- ☆ Partial shadow for a half cell in the bottom of the PV module.
- ☆ Partial shadow for a half cell in the middle of the PV module.
- ☆ Partial shadow for a half cell in the top of the PV module.
- ☆ Partial shadow for one cell of the PV module.
- ☆ Partial shadow for two cells of the PV module.

In each of these shadow patterns, the I-V and P-V measurements are taken simultaneously for two identical PV modules; one facing to the sun without shading and the other is partially shaded, for comparison the effect in the same operating conditions. The two PV modules are put in the same tilted surface with the same inclined angle.

The I-V and P-V measurements of the PV module under the specified partial shadow patterns are taken by using the simple and accurate I-V measuring circuit of photovoltaic modules based on an electronic load as shown in Figure 2.2 [19].

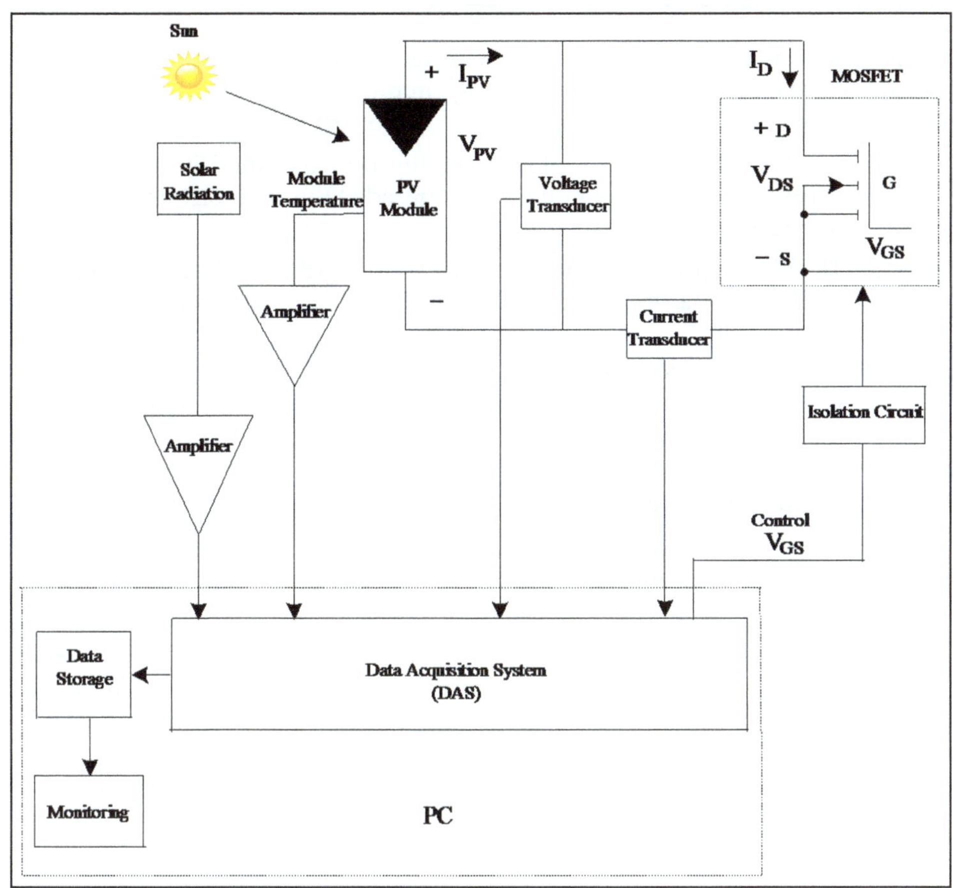

Figure 2.2: Block Diagram of the Electronic Circuit for Tracing I-V Characteristic [19].

As shown in Figure 2.1, the tracking of the PV module I-V and P-V curves is via electronic load using 520 W, N-type Metal Oxide Semiconductor Field Effect Transistor (MOSFET) with the required different sensors and transducers as well as the amplification and isolation circuits. Table 2.2 shows the electrical parameters of the electronic load.

Table 2.2: Electrical Parameters of the Electronic Load

Item	*Specifications*
Rated power	520 W
Operating voltage	500 V
Operating current	48 A
Operating temperature	Up to 150 °C

For each shadow patterns, the following electrical and physical parameters are measured and stored using the measuring circuit shown in Fig. 2 [19,20] for both modules at the same time under climates of Cairo, Egypt;

- ✰ Incident global solar radiation on the surface of the PV module using a Kipp-Zonen type pyranometer mounted at the same structure parallel to the PV module.
- ✰ The PV module surface temperature was recorded by the measuring circuit via K-type thermocouple placed at the centre of the back surface of the PV module.
- ✰ The PV module voltage can be tracked via a bipolar voltage transducer, while the current can be measured by the shown appropriate current transducer.

4. Result and Discussions

According to the shown shadow patterns, and the measured data from both PV modules (with and without shadow), the effect of partial shadow can be explained from the following results an discussions.

Figures 2.3 and 2.4 show the effect of partial shadow for a half cell in the bottom of the PV module on I-V and P-V characteristics of the PV module at solar radiation level of 975 W/m^2, respectively. From the figures, it is clear that the effect of half cell shaded directly affecting the power delivered by the PV module and the shape of the I-V curve has changed specially at the maximum power point. The loss of the PV module output power due to partial shadow of the half cell is about 7 per cent.

The same effect can be shown at any half cell of the PV module; bottom, middle or top, as shown in Figures 2.5 and 2.6 that represent I-V and P-V characteristics of the PV module for a bottom, middle and top half cell shaded (975 W/m^2), respectively.

Figures 2.7 and 2.8 show the effect of partial shadow for one and two adjacent cells of the PV module, respectively. From the figures it is clear that the same effect on the shape of the I-V and P-V curves as in case of half cell shaded. Simply shading one cell of the module causes a performance decrease; shading two causes performance to decrease further. The resultant power losses are 22 per cent and 41 per cent for one and two cells shaded, respectively. Also the open circuit voltage was highly affected in case of shaded two adjacent solar cells of the PV module.

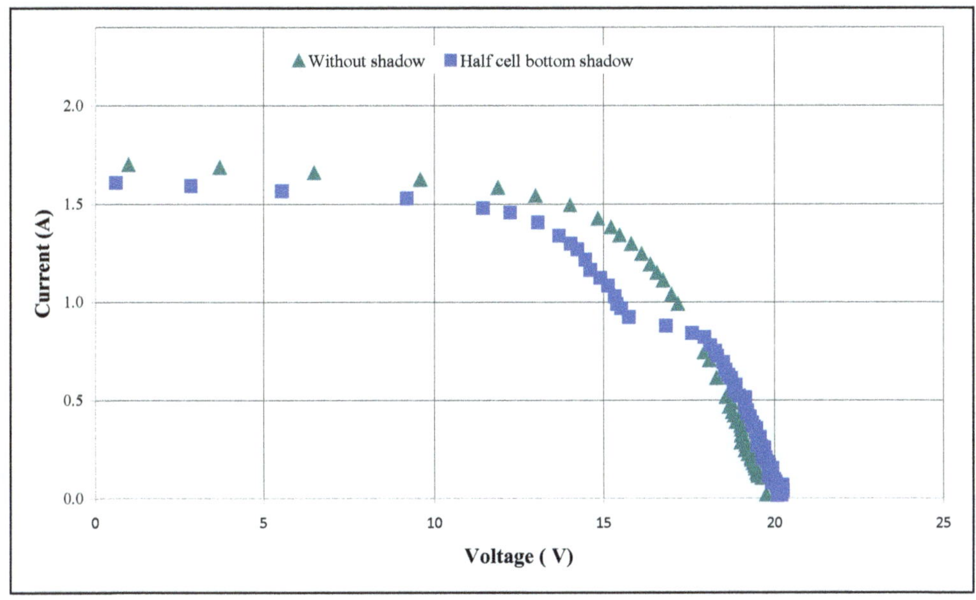

Figure 2.3: I-V Characteristics of the PV Module for a Bottom Half Cell Shaded.

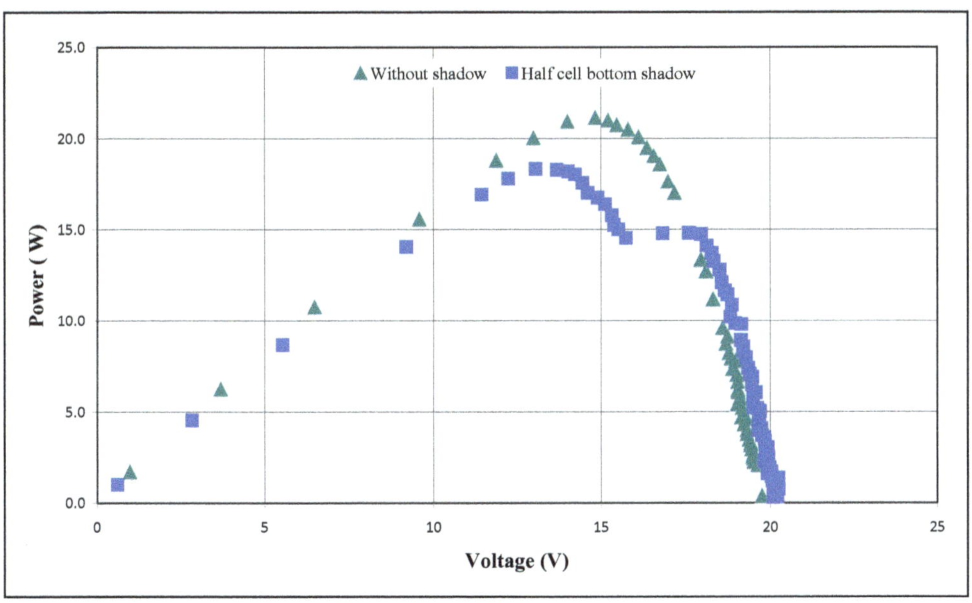

Figuer 2.4: P-V Characteristics of the PV Module for a Bottom Half Cell Shaded.

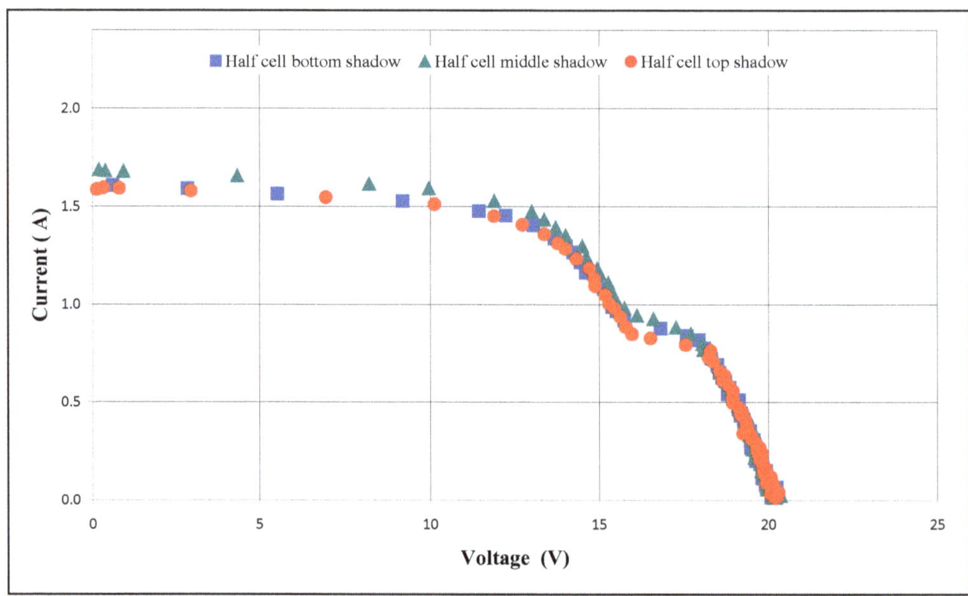

Figure 2.5: I-V Characteristics of the PV Module for a Bottom, Middleand Top Half Cell Shaded.

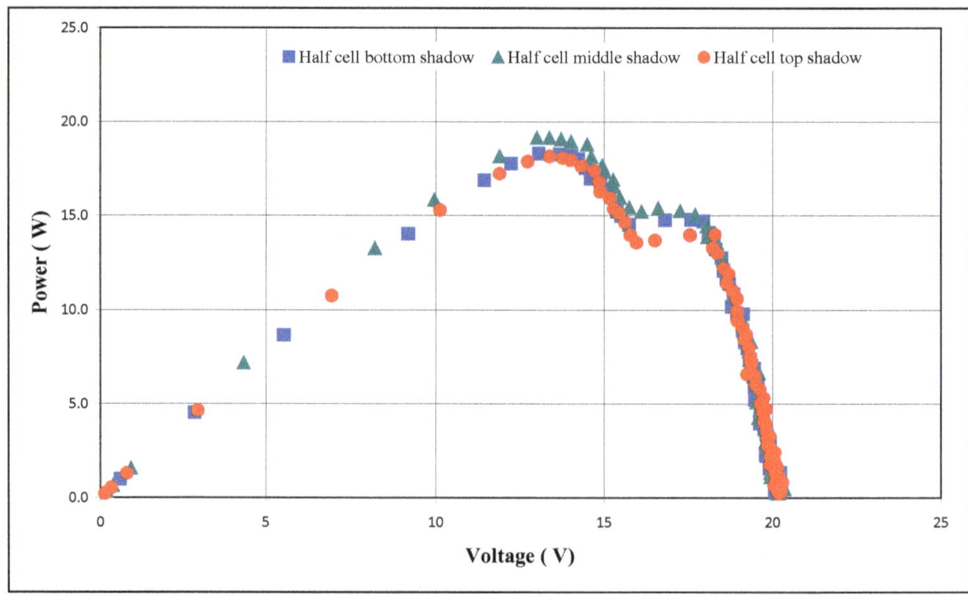

Figure 2.6. P-V Characteristics of the PV Module for a Bottom, Middleand Top Half Cell Shaded.

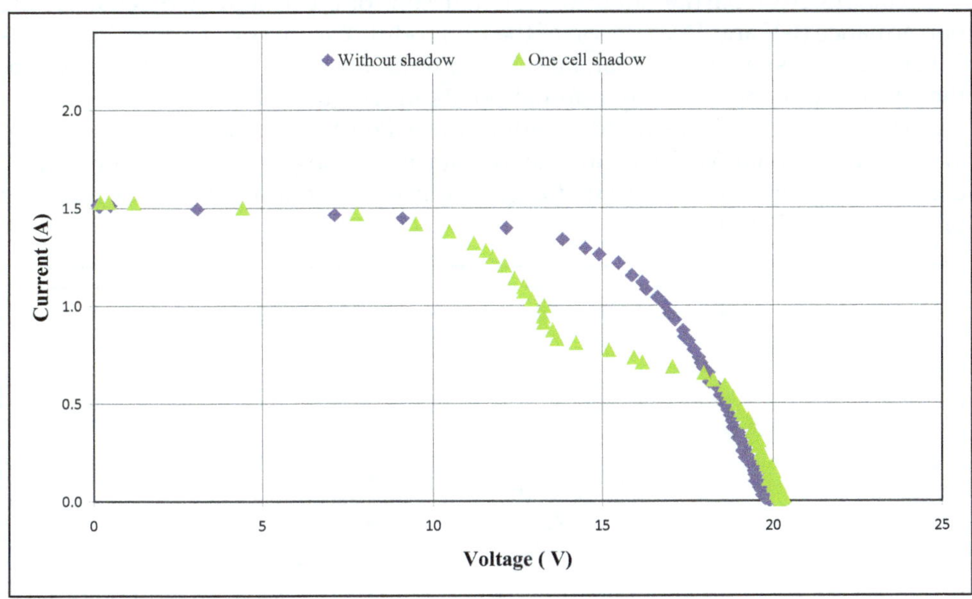

Figure 2.7: I-V Characteristics of the PV Module for One Cell Shaded.

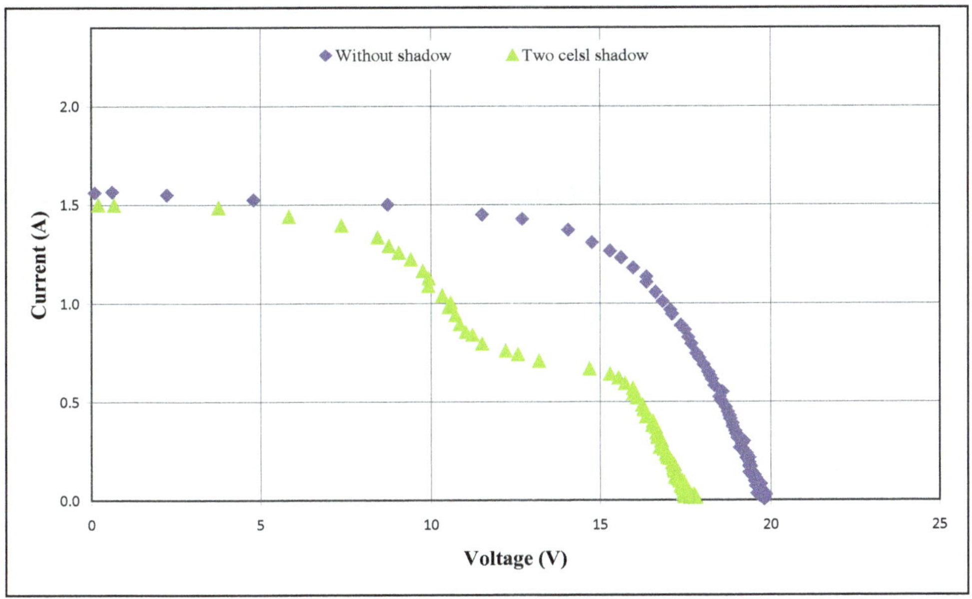

Figure 2.8: I-V Characteristics of the PV Module for Two Cell Shaded.

The effect of partial shadow (for half cell, one cell and two cells) on the performance (I-V and P-V curves) of the PV module can be summarized in Figures 2.9 and 2.10, respectively. It can be seen from the previous I-V and P-V curves of the PV module appear many optimal points in irregular shadow, which cause a big problem in case of using Maximum Power Point Tracking (MPPT) in the PV system. Traditional MPPT control algorithm is no longer available, new MPPT control algorithm must be studied in case of irregular shading that may affecting on the PV installation.

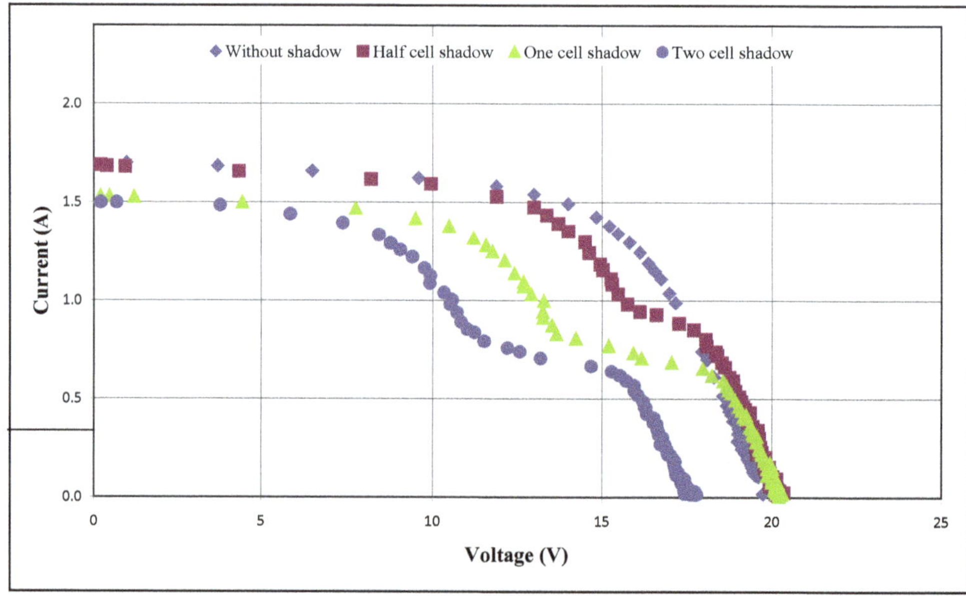

Figure 2.9: I-V Characteristics of the PV Module for Half Cell, One Cell and Two Cells Shaded.

In the selected PV module, all the cells are connected in series, thus the same current must flow through each cell. Partial shading causes a reduction of irradiation and thus a reduction of the current through the shaded cells. On the other hand, the un-shaded cells will force a larger current through the shaded cells. Since the current must be the same through the module, the difference will circulate through the shaded cells causing; heat dissipation leading to "hot spots" on the shaded cells and a reduction of the module output voltage. The latter translates in an inefficient operation of the module. In the long term, hot spots will reduce the lifespan of the module and must be avoided. A standard solution for this problem is the use of bypass diode protection. When the current increases above a certain threshold, the bypass diode becomes forward biased and provides a path for the current to circulate, thereby bypassing the shaded cells. Thus the bypass diode has been shown to improve the power produced and prevent the "hot spots" problem. A PV module will operate most efficiently if a bypass diode is connected in parallel

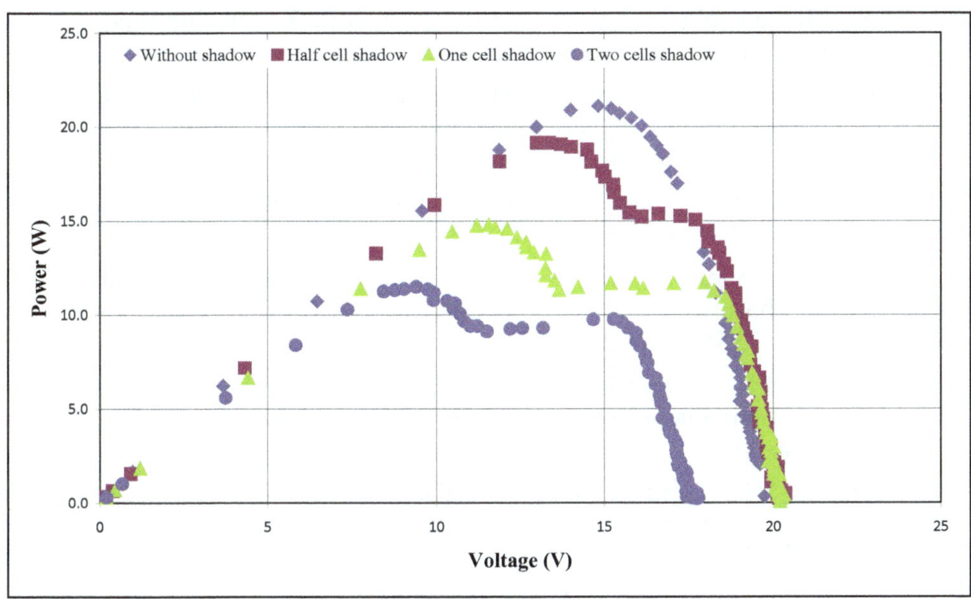

Figure 2.10: P-V Characteristics of the PV Module for Half Cell, One Cell and Two Cells Shaded.

with every cell. However, this is impractical from a design perspective. At least one bypass diode is added to protect the module and preserve the integrity of the array.

5. Conclusions

The effect of partial shading on the performance of a thin film PV module was investigated practically through the direct measurement of the I-V and P-V curves of the PV module under different partial shading patterns. The partial shading patterns are; half cell shaded in bottom, middle and top of the PV module, also for one and two cells. For accurate measurements, two identical modules (with and without shading) are placed at the same time on one mechanical structure for measurements. For all shading patterns, the shapes of the I-V and P-V curves are changed and have more than one maximum power point. This problem can inversely affect the system in case of using traditional maximum power point trackers. Also the power loss due shading are 7 per cent, 22 per cent and 41 per cent for shaded half cell, one cell and two adjacent cells of the PV module, respectively.

REFERENCES

1. Doty, G., Mc Cree, D., Doty, J. and Doty, F. Deployment prospects for proposed sustainable energy alternatives in 2020. *In:* International Proceedings of the ASME 20104th International Conference on Energy Sustainability, 2010, pp. 171–182.

2. REN21, Renewable 2012 global status report, 2012. Renewable energy policy network for the 21st century.
3. Fialho, L., Melicio, R., Mendes, V. M. F., Figueiredo, J.,and Collares-Pereira, X., 2014. Effect of shading on series solar modules: simulation and experimental results. Procedia Technology, 17, pp. 295-302.
4. Duffie, J.A. and Beckman, W.A., 2006. Solar Engineering of Thermal Processes. John Wiley and Sons, New Jersey, USA.
5. Patel, H. and Agarwal, V., 2008. MATLAB-based modeling to study the effects of partial shading on PV array characteristics. IEEE Transactions Energy Conversion, 23, pp. 302-310.
6. Quaschning, V. and Hanitsch, R., 1995. Numerical simulation of photovoltaic generators with shaded cells. Int. Proc. 30th Universities Power Engineering Conference, London, UK, pp. 583-589.
7. Woyte, A., Nijs, J. and Belmans, R., 2003. Partial shadowing of photovoltaic arrays with different system configurations: literature review and field test results. Solar Energy, 74, pp. 217-250.
8. Lua, F., Guoa, S., Walsha, T.M. and Aberlea, A.G., 2013. Improved PV module performance under partial shading conditions. Energy Procedia, 33, pp. 248-255.
9. Alonso-Garc ´a, M.C. and Ruiz, J.M., 2006. Analysis and modeling the reverse characteristic of photovoltaic cells. Solar Energy Materials and Solar Cells, 90, pp. 1105-1120.
10. Brecl, K. and Topi, M., 2011. Self-shading losses of fixed free-standing PV arrays. Renewable Energy, 36, pp. 3211-3216.
11. Karatepe, E., Boztepe, M. and Colak, M., 2007. Development of a suitable model for characterizing photovoltaic arrays with shaded solar cells. Solar Energy, 81, pp. 977-992.
12. Silvestre, S. and Chouder, A., 2008. Effects of shadowing on photovoltaic module performance. Progress in Photovoltaics. Research and Applications, 16, pp. 141-149.
13. Blaesser, G. and Munro, D., 1995. Guidelines for the assessment of photovoltaic plants, document C: initial and periodic tests on photovoltaic plants. Joint Research Centre of the European Communities, Ispra Establishment, Report EUR 16340 EN, Ispra, Italy.
14. Klise, G.T. and Stein, J.S., 2009. Models used to assess the performance of photovoltaic systems. Sandia National Laboratories, Report SAND.
15. Institute of Environmental Sciences (ISE), 2012. PVSYST V5.62, software for photovoltaic system", University of Geneva.
16. http://www.pvsyst.-com/en (accessed 19.12.12)
17. National Renewable Energy Laboratory (NREL), 2012. PVWatts V2. a performance calculator for grid-connected PV systems.

18. http://www.nrel.gov/rredc/pvwatts (accessed 19.12.12)
19. Martinez-Moreno, F., Munoz, J. and, Lorenzo, E., 2010. Experimental model to estimate shading losses on PV arrays. Solar Energy Materials and Solar Cells, 94, pp. 2298-2303.
20. Rodrigo, P., Ferna´ndez, E. F., Almonacid, F. and Pe´rez-Higueras, P.J., 2013. A simple accurate model for the calculation of shading power losses in photovoltaic generators. Solar Energy, 93, pp. 322-333.
21. El Shenawy, E.T., Esmail, O.N.A., Adel A. Elbaset and Hesham F.A. Hamed, 2014. Simple and accurate I-V measuring circuit for photovoltaic applications. International Journal of Engineering Research and Technology, 3(6), pp. 1141-1147.
22. Esmail, O.N.A., El Shenawy, E.T., Adel A. Elbaset and Hesham F. A. Hamed, 2013. Design and practical implementation of a data acquisition system for photovoltaic applications. Journal of Applied Sciences Research, 9(8), pp. 4856-4866.

Chapter 3

Current Status of Solar Power in India

Meenu Galyan

Centre for Science and Technology of the Non-aligned and Other Developing Countries (NAM S&T Centre), New Delhi, India
E-mail: meenu4sweet@gmail.com, meenugalyan07@gmail.com

The demand for energy is rapidly accelerating day by day beyond the limits of installable generation capacity in the whole world.India is ranked fifth in the electricity generation in the world. According to Ministry of Power, India's total installed capacity of 329,231MW out of which 2,20,576MW is from thermal, 44,614MW from hydro, 6,780MW from nuclear power stations and 57,260MW from Renewable Energy Sources [As on 30 June 2017 *Installed capacity in respect of RES (MNRE) as on 31 March 2017]. Thermal source contributes 67 per cent in the total capacity out of which 59.1 per cent by coal. Thus, Indian power sector is basically based on fossil fuels, about three-fifths of the country's power is generating by reserves of coal. Due to Limited fossil resources and environmental problems associated with them have emphasized the need for new sustainable energy supply options that use renewable energies.

Renewable energy sources and technologies provide solutions to the longstanding energy problems being faced by the developing countries like India. India being a tropical country has a tremendous scope of generating electricity from solar radiation. Therefore in India Solar energy is playing a pivotal role in compensating the electrical energy. Solar energy can be an important part of India's plan not only to add new capacity but also to increase energy security, address environmental concerns, and lead the massive market for renewable energy. Solar Thermal Electricity (STE) also known as Concentrating Solar Power (CSP) are emerging renewable energy technologies and can be developed as future potential option for electricity generation in India.

Sometimes however, their wide scale deployment has to face potential negative environmental implications. Major challenges of solar energy are its unavailability due to different weather condition and also storage issue.Further, the cost of electricity generated from solar is still expensive and also the power from renewable resources including solar is infirm power, large scale development of renewable resources did not take place and distribution utilities are also least interested to purchase power from renewable sources.

In present scenario, the solar energy based power generating systems can play a major role towards the fulfillment of energy demand in India. According to ministry of power, total cumulative power generation capacity of solar energy projects in India is about 9,235.24 MW (as on 31.01.2017).India is planning to produce 100 GW of solar power by 2022. Solar energy in India has experienced phenomenal growth in recent years due to both technological improvements resulting in cost reductions and government policies supportive of renewable energy development and utilization.

This paper deals with the availability, current status, strategies, perspectives, promotion policies and major achievements on solar power generation in India.

***Keywords**: Renewable energy, Solar energy, Challenges and policies, Power generation.*

1. Introduction

Electricity is the key factor for industrialization, urbanization, economic growth and improvement of quality of life in society (V. Khare *et al.*, 2013). India is ranked fifth in the electricity generation in the world. India has total installed capacity of 329,231MW out of which 67. per cent is from thermal, 13.6 per cent from hydro, 2.1 per cent from nuclear power stations and 17.4 per cent from Renewable Energy Sources [As on 30 June 2017 *Installed capacity in respect of RES (MNRE) as on 31 March 2017] (MoP, 2016-17). Although Over the years, Indian power sector has experienced a ten-time increased in its installed capacity–a jump from 30,000 MW in 1981 to over 329,231 MW page9by 30 June 2017 but still there is a huge gap in generation and demand in India hence need to be establish more generation plants preferably to be come from renewable sources by governmental as well as various private participation (MoP, 2016-17). Thus, Indian power sector is basically based on fossil fuels, about three-fifths of the country's power is generating by reserves of coal. The huge consumption of fossil fuels has caused visible damage to the environment in various forms. Every year human activity dumps roughly 8 billion metric tonnes of carbon into the atmosphere, 6.5 billion tonnes from fossil fuels and 1.5 billion from deforestation [Sood YR, Padhy NP, Gupta HO, 2002]. India also has followed the global change in power sector by establishment of the Regulatory Commissions in 1998 under the Electricity Regulatory Commissions Act 1998 (Central Law) to promote competition, efficiency and economy in the activities of the electricity industry and applied restructuring to Orissa state electricity board firstly and after that too many other states [Singh A, 2006].

The rapid and accelerated growth in energy demand and energy consumption, particularly in the past several decades, has raised fears of exhausting the globes reserves of petroleum and other resources in the future. To overcome the problems associated with the generation of electricity from fossil fuels, renewable energy sources can participate in the energy mix.In the near future, due to the global population growth and industrialization, the demand for electric energy is expected to increase rapidly [C. Vidya, 2017]. World's net electricity generation is expected to rise from 23.5 trillion KWh in 2015 to 26.6 trillion KWh in 2020 and 36.2 trillion KWh in 2035 [N.D.Mani *et al.*, 2016].

One of the ways to sustainable growth is to generate electricity through solar energy which is cleaner and promising [K. Kapoor *et al.*, 2014]. Solar has the greatest energy potential among the other sources of renewable energy and the amount of energy that earth receives, as assessed by IPCC, theoretically if only a small fraction of this form of energy could be used it can meet our current needs [Singhal SK and Varun, 2006] or in other words if we can use only 5 percentage of this energy, it will be 50 times what the World requires [Rai GD., 2000].

2. Overview of Solar Power in India

Power sector plays a vital role in the growth of Indian economy and it is growing at rapid pace [Annual Report 2016-17]. India is among the leading countries having good Direct Normal Irradiance (DNI is solar radiation that comes in a straight line from the direction of the sun at its current position in the sky), which depends on the geographic location, earth – sun movement, tilt of Earth rotational axis and atmospheric attenuation due to suspended particles [K. Kapoor *et al.*, 2014]. India is estimated to have huge potential for solar energy which is about 5000 trillion kWh per year (*i.e.* ~600 TW), far more than its current total primary energy consumption (Rachit.S *et al.*, 2015,). The solar radiation incident over India is equal to 4 – 7 kWh per square meter per day [Kumar A and Kumar K, 2010] with an annual radiation ranging from1200 – 2300 kWh per square meter [MNRE, Anuual Report, 2006]. It has an average of 250 – 300 clear sunny days [Kumar A and Kumar K, 2010] and 2300 – 3200 hours of sunshine per year [Sharma NK and Tiwari PK, 2011].

Solar radiation levels in different parts of the country are given in Figure 3.1. It can be observed that although the highest annual global radiation is received in Rajasthan, northern Gujarat and parts of Ladakh region, the parts of Andhra Pradesh, Maharashtra, and Madhya Pradesh also receive fairly large amount of radiation as compared to many parts of theworld especially Japan, Europe and the US where development and deployment of solar technologies is maximum [Garud S, Purohit I.]. Therefore, there is unlimited scope for solar electricity to replace all fossil fuel energy requirements (natural gas, coal, lignite and crude oil) if all the marginally productive lands are occupied by solar power plants in future. The solar power potential of India can meet perennially to cater per capita energy consumption at par with USA/Japan for the peak population in its demographic transition. (Vinod.KG 2015).

According to ministry of power, total cumulative power generation capacity of solar energy projects in India is about 9,235.24 MW (as on 31.01.2017) with generation mix of Thermal, Hydro, Renewable and Nuclear.The data of total installed power generation capacity in India by the ministry of power in government of India as on 30 June 2017 were shown in Table 3.1 and also shown in Figure 3.2 and the sector-wise installed power were also tabulated in Table 3.2.

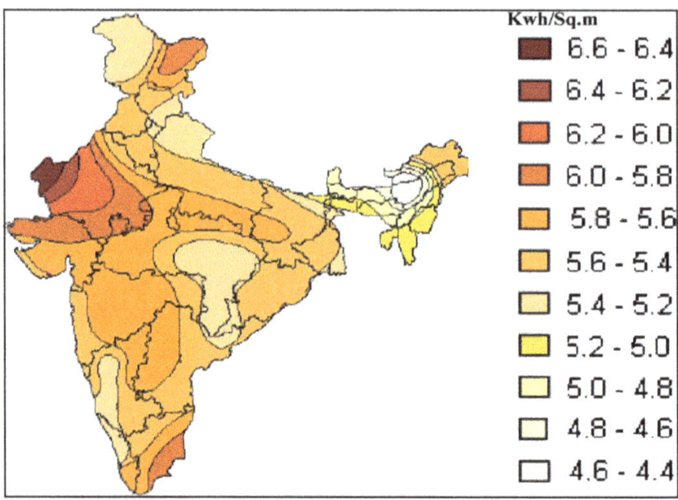

Figure 3.1: Solar Radiation on India.

Table 3.1: Total Installed Power Generation Capacity of India

Fuel	***MW (31 Dec 2015)***	***MW (30 Aug 2016)***	***MW (30 Jun 2017)***
Coal	1775238 (64 per cent)	186593 (61.12 per cent)	194553 (59.1 per cent)
Gas	24509 (8 per cent)	25057 (8.18 per cent)	25185 (7.6 per cent)
Diesel	994 (0.345 per cent)	919 (0.30 per cent)	838 (0.3 per cent)
Nuclear	5780 (2 per cent)	5780 (1.89 per cent)	6780 (2.1 per cent)
Hydro	42663 (14 per cent)	42663 (14 per cent)	44614 (13.6 per cent)
RES (MNRE)	38822 (12 per cent)	44237 (14.44 per cent)	57260 (17.4 per cent)
Total	288005	305554	**329231**

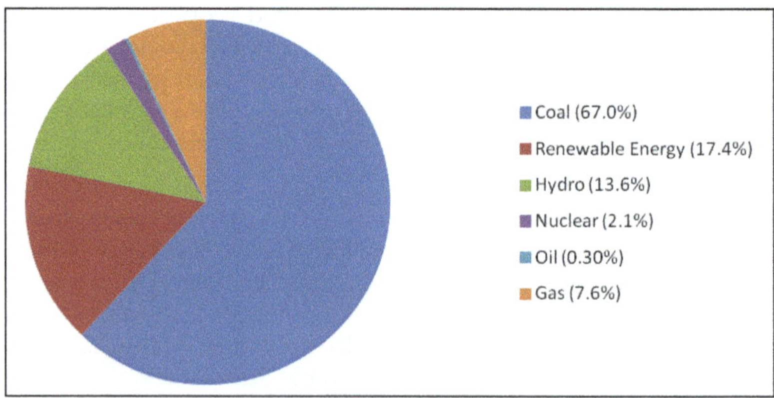

Figure 3.2: Total Installed Power Generation Capacity of India (As on 30 June 2017).

Table 3.2: Sector-wise Total Installed Power Generation Capacity of India

Sector	MW (31 Dec 2015)	MW (30Aug 2016)	MW (30 Jun 2017)
State Sector	97951 (34 per cent)	101946 (33 per cent)	103868 (31.55 per cent)
Central Sector	74807 (26 per cent)	76312 (25 per cent)	81622 (24.79 per cent)
Private Sector	115248 (40 per cent)	127296 (42 per cent)	143740 (43.66 per cent)
Total	**288005**	**305554**	**329231**

Current Status

India's cumulative solar capacity stands at 13114.85 MW as of 30 June 2017. The country installed an impressive 5,525.98 MW in 2016-17 according to MNRE. In contrast, India had a cumulative installed solar capacity of 6,762.85 MW at the end of March 2016 and had installed 3,010 MW in 2015-16. The State wise total cumulative capacity commissioning status of solar power projects in India as on 31st January 2017 were tabulated in Table 3.3. (MNRE Jan 2017). According to state wise installed solar power projects in total Installed capacity as on 31.Jan.2017 has reaches 9235.24 MW.

Table 3.3: State-wise Total Cumulative Capacity Commissioning Status of Solar Power Projects in India as on 31st January 2017

Commissioning Status of Solar Power Projects as on 31-01-2017				
Sl.No.	**State/UT**	**Total Cumulative Capacity till 31-03-16 (MW)**	**Capacity Commissioned in 2016-17 till 31-01-17 (MW)**	**Total Cumulative Capacity till 31-01-17 (MW)**
1	Andaman and Nicobar	5.10	0.30	5.40
2	Andhra Pradesh	572.97	406.68	979.65
3	Arunachal Pradesh	0.27	0.00	0.27
4	Assam	0.00	11.18	11.18
5	Bihar	5.10	90.81	95.91
6	Chandigarh	6.81	9.40	16.20
7	Chhattisgarh	93.58	41.61	135.19
8	Dadar and Nagar	0.00	0.60	0.60
9	Daman and Diu	4.00	0.00	4.00
10	Delhi	14.28	24.50	38.78
11	Goa	0.00	0.05	0.05
12	Gujarat	1119.17	40.58	1159.76
13	Haryana	15.39	57.88	73.27
14	Himachal Pradesh	0.20	0.13	0.33
15	J&K	1.00	0.00	1.00
16	Jharkhand	16.19	1.33	17.51
17	Karnataka	145.46	196.46	341.93
18	Kerala	13.05	2.81	15.86

Commissioning Status of Solar Power Projects as on 31-01-2017				
Sl.No.	**State/UT**	**Total Cumulative Capacity till 31-03-16 (MW)**	**Capacity Commissioned in 2016-17 till 31-01-17 (MW)**	**Total Cumulative Capacity till 31-01-17 (MW)**
19	Lakshadweep	0.75	0.00	0.75
20	Madhya Pradesh	776.37	73.98	850.35
21	Maharashtra	385.76	44.70	430.46
22	Manipur	0.00	0.01	0.01
23	Meghlya	0.00	0.01	0.01
24	Mizoram	0.10	0.00	0.10
25	Nagaland	0.00	0.50	0.50
26	Odisha	66.92	10.72	77.64
27	Puducherry	0.03	0.00	0.03
28	Punjab	405.06	187.29	592.35
29	Rajasthan	1269.93	47.71	1317.64
30	Sikkim	0.00	0.01	0.01
31	Tamil Nadu	1061.82	529.15	1590.97
32	Telangana	527.84	545.57	1073.41
33	Tripura	5.00	0.02	5.02
34	Uttar Pradesh	143.50	125.76	269.26
35	Uttarakhand	41.15	3.95	45.10
36	West Bengal	7.77	15.30	23.07
37	Other/MoR/PSU	58.31	3.39	61.70
	TOTAL	6762.85	2472.39	**9235.24**

3. Indian Government Incentives and Support

Since 2000, there are various initiatives taken by Government of India to encourage solar energy in the country namely, Electricity Act 2003, National Electricity Policy 2005, Tariff Policy 2006 and its amendment in 2011, National Action Plan on Climate Change 2008, Semiconductor Policy 2007 and Karnataka Semi- conductor Policy 2010.

Electricity Act, 2003

It outlines several enabling provisions to accelerate the development of Renewable Energy based generation such as **[CERC,Electricity Act, 2003]**:

- ☆ (Section 3): it states that "under the National Electricity Policy Plan, Central Government in consultation with the State Government will develop policy and plans for development of power system based on renewable sources of energy and other conventional energy sources."
- ☆ (Section 61(h)): it states that "the appropriate commission shall specify the terms and conditions for determination of tariff and in doing so, shall promote generation of electricity from renewable sources of energy in

their area of jurisdiction."

- ☆ (Section 66): "Appropriate Commission shall endeavor to promote the development of market (including trading) in power in such a manner as may be specified and shall be guided by National Electricity Policy in Section 3 in this regard."
- ☆ (Section 86(1) (e)): "The State Electricity Regulatory Commissions to specify minimum Purchase Obligation from renewable sources of energy out of total consumption of energy (a percentage of the total consumption of electricity in the area of a power distribution company)."
- ☆ (Section 86(1)(e)) reads as "promote cogeneration and generation of electricity from renewable sources of energy by providing suitable measures for connectivity with the grid sale of electricity to any person, also specify, for purchase of electricity from such sources, a percentage of the total consumption of electricity in the area of a distribution licensee"

National Electricity Policy (February 2005)

Specify that, "share of electricity from renewable sources would need to be increased progressively, purchase for distribution companies will rely on competitive bidding basis. Commission needs to determine an appropriate differential tariff to promote these technologies, as they will take some time to compete with conventional sources on basis prices" [National electricity policy, 2005].

Tariff Policy (January 2006)

Government of India announced the Tariff Policy on 6th January 2006 in accordance with Section 3 of Electricity Act 2003,it states that appropriate Commission (State Electricity Regulatory Commission) shall fix Renewable Purchase Obligation (RPO) and shall also fix the technology specific tariff for different renewable sources of energy. The Tariff Policy provided that, initially the appropriate Commission is to fix preferential tariffs for distribution utility to procure Renewable Energy (RE), in future, distribution utility to procure RE technology through competitive bidding with in suppliers offering same type of RE. In long term, RE technologies are expected to compete with all other sources in terms of cost [Tariff policy, 2006].

The Amendment in Tariff Policy 2011 [Amendment in Tariff Policy, 2011]

- ☆ SERCs shall fix separate RPO for purchase of energy by the Obligated Entities from Solar Energy Source
- ☆ Solar RPO to go up to 0.25 per cent by the end of 2012–13 further up to 3 per cent by2022
- ☆ Purchase of energy from non-conventional sources of energy to take place more or less in same proportion across different States _ Renewable Energy Certificate (REC) Mechanism may be one of the mechanisms to achieve such target

National Action Plan on Climate Change (NAPCC) - 2008

It earmarks the National level target for RE Purchase from 5 per cent of total grid purchase in 2010 to 15 per cent by 2020 by increase of 1 per cent each year for next 10 years from 2010. Different SERCs of respective states may set higher target independently. It also authorizes appropriate authorities to issue certificates that procure RE in excess of the national grid, which may be tradable, to enable utilities falling short to meet their obligation of renewable energy purchase (RPO) [GOI, NAPCC, 2008].

Semiconductor Policy 2007

This policy was announced by Government of India to draw investments in semiconductor fabrication by providing special incentives for manufacturing of all semiconductors and solar photovoltaic cells [GOI, Semiconductor policy, 2007].

Karnataka Semiconductor Policy 2010

Government Initiatives

The Government of India has identified power sector as a key sector of focus so as to promote sustained industrial growth. Some initiatives by the Government of India to boost the Indian power sector:

- ☆ The Union Cabinet, Government of India has given its ex-post facto approval for signing of a Memorandum of Understanding (MoU) on Renewable Energy between India and Portugal, which will help strengthen the bilateral cooperation between the two countries.
- ☆ The Ministry of New and Renewable Energy plans to introduce a fixed-cost component to the tariff for electricity generated from renewable energy sources like solar or wind, in a bid to promote a green economy.
- ☆ The Union Cabinet has approved the ratification of International Solar Alliance's (ISA) framework agreement by India, which will provide India a platform to showcase its solar programmes, and put it in a leadership role in climate and renewable energy issues globally.
- ☆ The Government of India plans to rationalise various categories of electricity consumers across states, which is expected to bring transparency and efficiency in billing, improve tariff collection and improve the health of distribution companies in the country.
- ☆ The Government of India plans to set up a US$ 400 million fund, sourced from The World Bank, which would be used to protect renewable energy producers from payment delays by power distribution firms, while at the same time protecting the distribution firms from the shrinking market for conventional grid-connected power, caused by wider adoption of roof-top solar power generation.
- ☆ The Ministry of Power plans to set up two funds of US$ 1 billion each, which would give investment support for stressed power assets and renewable energy projects in the country.

- ☆ Mr Piyush Goyal, Minister of State with Independent Charge for Power, Coal, New and Renewable Energy and Mines, launched an online portal for star rating of mines, which will bring all mines to adopt sustainable practices, and thereby ensure compliance of environmental protection and social responsibility by the mining sector.
- ☆ The Ministry of New and Renewable Energy (MNRE), which provides 30 per cent subsidy to most solar powered items such as solar lamps and solar heating systems, has further extended its subsidy scheme to solar-powered refrigeration units with a view to boost the use of solar-powered cold storages.
- ☆ Mr Piyush Goyal, Minister of State with Independent Charge for Power, Coal, New and Renewable Energy and Mines, inaugurated the Tarang (Transmission App for Real Time Monitoring and Growth) mobile app and web portal for electronic bidding for transmission projects, which is expected to enhance ease, accountability, transparency, and boost investor confidence in power transmission sector.
- ☆ The Ministry of Shipping plans to install 160.64 MW of solar and wind based power systems at all the major ports across the country by 2017, thereby promoting the use of renewable energy sources and giving a fillip to government's Green Port Initiative.
- ☆ The Government of India and the Government of the United Kingdom have signed an agreement to work together in the fields of Solar Energy and Nano Material Research, which is expected to yield high quality and high impact research outputs having industrial relevance, targeted towards addressing societal needs.
- ☆ The Government of India plans to start as many as 10,000 solar, wind and biomass power projects in next five years, with an average capacity of 50 kilowatt per project, thereby adding 500 megawatt to the total installed capacity.
- ☆ Government of India has asked states to prepare action plans with year-wise targets to introduce renewable energy technologies and install solar rooftop panels so that the states complement government's works to achieve 175 GW of renewable power by 2022.
- ☆ The Government of India announced a massive renewable power production target of 175,000 MW by 2022; this comprises generation of 100,000 MW from solar power, 60,000 MW from wind energy, 10,000 MW from biomass, and 5,000 MW from small hydro power projects.

4. Barriers and Challenges on Solar Energy in India

Technical Barriers

The technology is surrounded by many technical issues and uncertainties, such as storage issues, the risk associated with technology and its related system, the inadequacy and reliability of the DNI data available and many others have always been a matter of debate. Due to these concerns it has become difficult for solar to be competitive with other more established sources of energy.

Policy and Regulatory Barriers

To grow consistently solar energy sector requires supportive policies and encouraging regulations. The investors show reluctance and uncertainty if they perceive high risk in the sector, which can only be assured by strong and attractive policies for developing a market. The lack in clarity of policies and regulations can adversely affect the long term development planning of a country.

Socio-economic Challenges

The Socio-Economic challenges have a strong impact on the development of solar energy, as it can leads to less adoption and acceptance of the technology. It is a hard fact that the technology requires huge investment which alone cannot be fulfilled through sovereign funds, so there is need to draw private investments, which can only be attracted when there are favorable incentives to invest in the technology. There are concerns to availability of affordable credit, as the market is not mature enough to help attract investments for development of the sector.

Institutional Challenges

The government organizations and other institutions play important role in developing and promoting a technology in a country. In India, we have seen organizations like IREDA, CASE, SAP *etc.*, being setup to promote renewable, but impact of these initiatives have not been promising. In fact a well-functioning institution can alone solve many challenges for solar to grow steadily.

5. Conclusions

In this paper, we have discussed about the current status of solar energy in India. The Government of India is taking a number of steps and initiatives like 10-year tax exemption for solar energy projects, *etc.*, in order to achieve India's ambitious renewable energy targets of adding 175 GW of renewable energy, including addition of 100 GW of solar power, by the year 2022. This discussion shows that the status of solar energy is satisfactory in India but some extra effort is required for betterment of solar source. In spite of reduction of the cost of solar power, it is expensive source of power compared with conventional sources. It is very important to support and subsidize the solar power till it can compete with the conventional sources. The step of Indian government to increases the target is a very good to become India as one of the most solar powered countries in the world. Such types of steps will be required in the future.

REFERENCES

1. N.K. Sharma *et al.*, Solar energy in India: Strategies, policies, perspectives and future potential. Renewable and Sustainable Energy Reviews *16 (2012) 933–941*
2. V. Khare *et al.*, Status of solar wind renewable energy in India. Renewable and Sustainable Energy Reviews 27 (2013) 1–10

3. Ministry of Power (MoP), 2016-17. Load Generation Balance Report 2016-17 – Central Electricity Authority.
4. Sood YR, Padhy NP, Gupta HO. Wheeling of power under deregulated environment of power system-a bibliographical survey. IEEE Trans Power Syst 2002;17(3):870–8.
5. Singh A. Power sector reform in India: current issues and prospectus. Energy Policy 2006;34:2480–90.
6. K. Kapoor *et al.*, Evolution of solar energy in India: A review. RenewableandSustainableEnergyReviews40(2014)475–487.
7. Kumar A, Kumar K. Renewable energy in India: current status and future potentials. Renew Sustain Energy Rev 2010; vol. 14:24 34–42.
8. MNRE. Annual Report, New Delhi; 2006.
9. Sharma NK, Tiwari PK. Solar Energy in India: strategies, policies, perspectives and future potential. Renew Sustain Energy Rev 2011; vol. 16:933–41.
10. Garud S, Purohit I. Making solar thermal power generation in India a reality – overview of technologies, opportunities and challenges, The Energy and Resources Institute (TERI), Darbari Seth Block, IHC Complex, Lodhi Road, New Delhi 110003, India.
11. C. Vidya, P. M. Anbarasan and K. M. Prabu. Energy Crisis and Recent Technological Development in India. Journal of Energy Technologies and Policy. ISSN 2224-3232 (Paper). ISSN 2225-0573 (Online) Vol.7, No.1, 2017.
12. N.D. Mani and Penmaya Ningshen, (February 2016), Mechanical Engineering: An International Journal (MEIJ), Vol. 3, No. 1.
13. Atul S, (30 May 2016), A comprehensive study of solar power in India and World. Renewable and Sustainable Energy. Retrieved.
14. Rachit S., and Vinod.K.G. (2015), Journal of Engineering and Technology, ISSN: 2319-9873.
15. Singhal SK, Varun, Devolopment of non-conventional energy sources," 24 March 2006. [Online]. Available: http://www.energymanagertraining.com/Journal/24032006/Developmentofnonconventionalenergysources.pdf. [Accessed 12 April 2012].
16. Rai GD. Non Conventional Energy Sources. 4th ed. New Delhi, India: Khanna Publishers; 2000.
17. Government of India (GOI), Semiconductor Policy of India 2007; 2007. [accessed 15.06.13].
18. Government of India, National Action Plan on Climate Change. 2008.
19. PIB, Amendment in Tariff Policy, New Delhi: Press Information Bureau; 2011.
20. Ministry of Power, Tariff Policy, Ministry of Power, New Delhi; 2006.
21. Government of India, National Electricity Policy; 2014.
22. CERC, Electricity Act 2003; 2014.

Chapter 4

Optimization of Hydrogented Amorphous/Nanocrystalline $Si_{1-X}Ge_x$ Thin Films by Pulsed Plasma CVD

*Ayana Bhaduri[1] * and Partha Chaudhuri[2]*

[1]Department of Applied Physics,
Amity University Gurgaon, Gurgaon – 122 413, Haryana, India
[2]Centre of Excellence for Green Energy and Sensor Systems,
Indian Institute of Engineering Science and Technology, Shibpur,
Howrah – 711 103, West Bengal, India
E-mail: abhaduri@ggn.amity.edu, ayana.bhaduri@gmail.com

Incorporation of the particles formed in the plasma into the amorphous silicon based films and devices may have drastic effect on their performance. We will summarize the findings our works on a-$Si_{1-x}Ge_x$:H that intelligent control of the particle size and number density by superimposion of a Square Wave Pulse Modulation (SWPM) on the 13.56MHz rf PECVD may also sometimes give improved transport and optoelectronic properties of the material.

Keywords: *Amorphous hydrogenated silicon germanium, Nanocrystalline silicon germanium, PECVD, Square wave pulse modulation, Photoconductivity, Mobility-lifetime.*

1. Introduction

Hydrogenated amorphous silicon germanium is widely used for a number of solid state electronic devices, such as solar cells [1], solid state photosensors, photodetectors [2] and thin film transistor for liquid crystal displays, photoreceptors, microelectronic applications [3] *etc.* [4,5]. The incorporation of germanium into amorphous silicon leads to a reduction in band gap [6,7] and in enhancement in the absorption coefficient at longer wavelengths [8,9].The extensive applications of a-SiGe:H is due to its several unique features: 1> it has a very high optical absorption coefficient (>105 cm^{-1}) over the majority of visible spectrum results in extremely thin film device possible; 2> low temperature deposition process is applicable; 3> the tunable bandgap of 1.0~1.7eV lies near the energy at which high solar energy conversion efficiencies are expected; 4> the materials is easy to dope both p-type and n-type using boron or phosphorous respectively; 5> the electronic properties of electrons and holes are adequate for many device applications.

As hydrogenated amorphous silicon germanium (a-SiGe:H) is a low band gap semiconductor that could be used as active material in thin film silicon based solar cells. By controlling the Si/Ge ratio in hydrogenated amorphous silicon–germanium (a-Si_1-$_xGe_x$:H) alloy films, the band gap may be varied continuously from1.75eV(a-Si:H) to below 1eV (a-Ge:H). The direct band gap of the alloy films allows for the use of a very thin absorber layer and much shorter deposition times compared to microcrystalline silicon (μc Si:H). In high efficiency amorphous silicon-based tandem solar cells, amorphous silicon–germanium (a-Si_1-$_xGe_x$:H) alloy material is used in the lower intrinsic layers in order to absorb the longer wavelength region of the visible solar spectrum.[10]

In the past, United Solar Ovonic has developed a-SiGe:H in triple junction solar cells [11]. Unfortunately, the high defect density of low band gap a-SiGe:H films deposited by plasma enhanced chemical vapor deposition (PECVD), the occurrence of light induced degradation of devices with relatively thick absorber layers in particular and the high costs of Ge-containing source gasses, band-gap profiling in the absorber layer [12] have thus far prevented a-SiGe:H in thin film silicon based solar cells to be viable.

The nanostructured Si:H and nanostructured SiGe:H usually referred as polymorphous hydrogenated silicon (pm-Si:H) and polymorphous hydrogenated silicon germanium (pm-SiGe:H) respectively are promising materials for application in single and multijunction thin film solar cells [11, 13]. Incorporation of the nanoparticles (size range 2-10nm) formed in the plasma into the deposited layers have strong influence on their performance [14,15]. The main idea of this study is to regulate the incorporation of these fine particles into the a-SiGe:H films by applying square wave pulse modulation (SWPM) of the rf amplitude of conventional 13.56 MHz RF PECVD and observe the subsequent changes in the films structural and optoelectronic properties.

2. Experimenal Details

Using SiH_4-GeH_4 mixture as source gas with high H_2 dilution, two series of samples were deposited. The series named #DC, we kept the total pulse time period fixed at 737 μsec while gradually changed the "*duty cycle*" (DC) defined by,

$$DC = \frac{T_{on}}{T_{on} + T_{off}} \times 100\% \tag{1}$$

from DC=100 per cent (continuous mode) to 50 per cent by controlling the "on" and "off" times. Other deposition parameters, *viz.*, deposition temperature, T_{dep} =250 C, pressure, P = 66.7 Pa and rf power density, P_{rf} = 197 mW/cm^2 were kept same for both the series. The gas flows were 2.8 sccm (sccm is the cubic centimeter of gas flow per minute at STP) of SiH_4, 0.8 sccm of GeH_4 and 95 sccm of H_2 for the series.

The film microstructures linked to the incorporation of the nanostructured particles from the plasma were investigated by high-resolution transmission electron microscope (JEOL 2010). The particle density was counted from the micrograph with standard software "IMAGE-J". For compositional analysis of the nanoparticles the energy dispersive x - ray spectra (EDS) were collected in-situ during TEM imaging. Surface roughness of the various layers was measured by atomic force microscope (Vecco diCPII) using "Proscan" software. Optoelectronic properties like photo-sensitivity and the sub-band gap absorption of these films were measured and correlated with the microstructural data. Bandgap value (E_g) by means of ultraviolet and visible light (UV-vis spectroscopy), the ambipolar diffusion length (L_d) measured from Steady State Photocarrier Grating (SSPG) technique, and the activation energy (E_a) as well as the mobility-lifetime product ($\mu\tau$) as deduced from dark and photoconductivity measurements, respectively. Measurements of L_d, E_a, and $\mu\tau$ were performed in the as-deposited state of the films after light soaking (LS) for 36 h at 80 C under a 300 mW/cm^2 red light and after annealing of the light-soaked films at 190 C for 12 h in the dark.

3. Results

3.1. Optoelectronic Properties

Evolution of the ambipolar diffusion length L_d and $\mu\tau$product with the duty cycle, is shown below.The films being in the as deposited (As Dep.), Light-Soaked (LS) or annealed states, for the #DC series films.

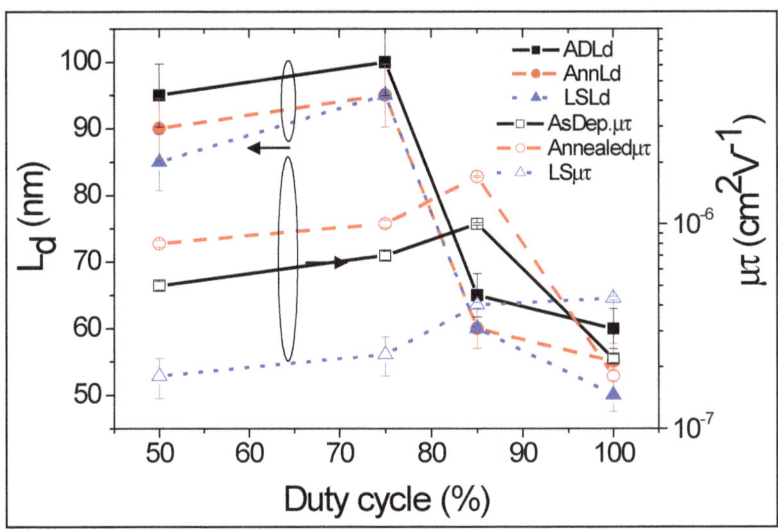

Figure 4.1: Light Induced Degradation of Ambipolar Diffusion length (L_d) and Mobility Lifetime Product (μt).

From the Figure 4.1, we could observed that on light soaking, the L_d values degrade less for DC=75 per cent. Moreover, after 100Hours of accelerated light soaking, L_d value reduces to 95 nm from 100 nm at DC=75 per cent. So only 5 per cent degradation in L_d value. We would like to point out that with band gap of 1.45 eV and Ge content around 60 per cent, $L_d$value of 100 nm is the highest obtained value ever reported for a-$Si_{1-x}Ge_x$:H thin films. [16, 17, 18].

3.2. Structural Study (by HRTEM and AFM)

We have shown the HRTEM fringe patterns of a-SiGe:H for 75 per cent DC in Figure 4.2.

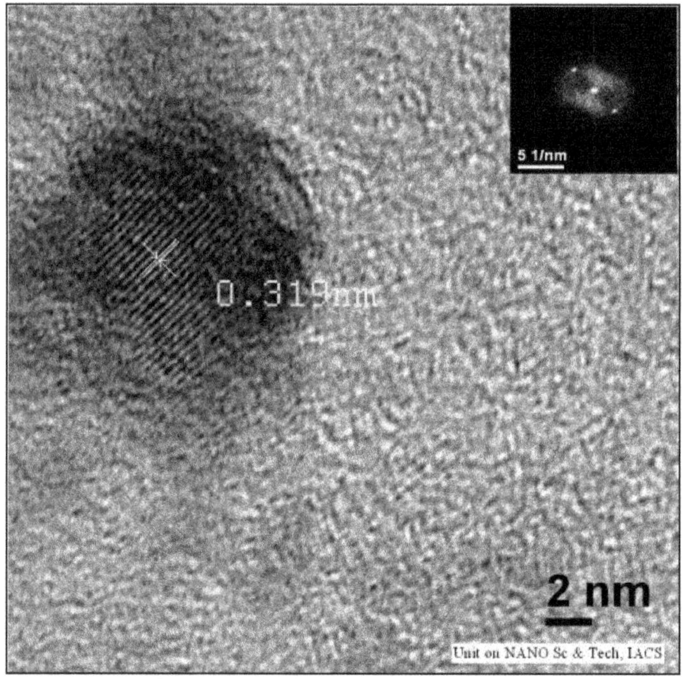

Figure 4.2: HRTEM Micrograph of a-SiGe: H for 75 per cent DC.

The micrograph depicts a two-phase material where isolated nanocrystallites and/or nanoclusters are embedded in an amorphous tissue. The lattice spacing 0.319 nm is corresponds to $Si_{0.5}Ge_{0.5}$(111) plane.

From the AFM micrograph, the embedded nanocrystallites/fine particle of isotropic shape and nearly uniform size distribution at DC 75 per cent films is observed. [19]

Statistical analysis of the surface roughness (R_{RMS}) from the AFM study has been done for different samples from duty cycle continuous mode to 50 per cent. It has been observed that there is a significant increase in R_{RMS} from 1 to 12 nm with the lowering of the duty cycle from 100 per cent (continuous mode) to 50 per cent indicating presence of nanocrystallites.

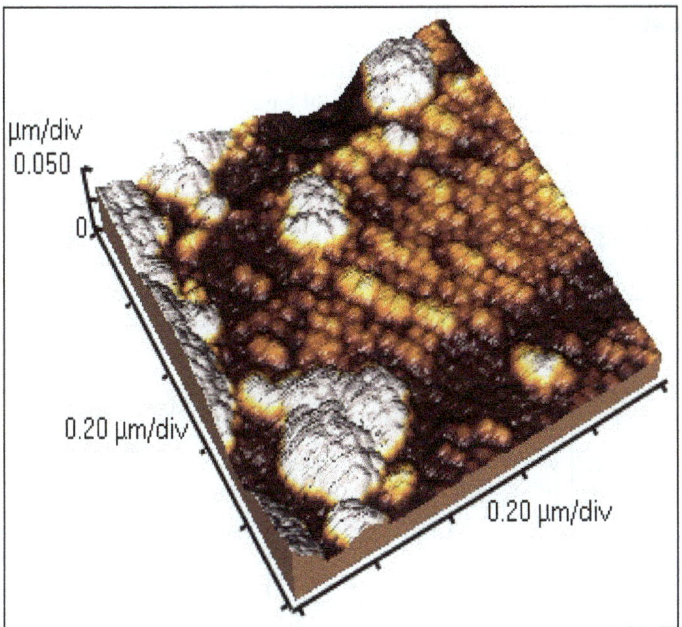

Figure 4.3: AFM 3D Surface Topography Image of DC = 75 per cent.

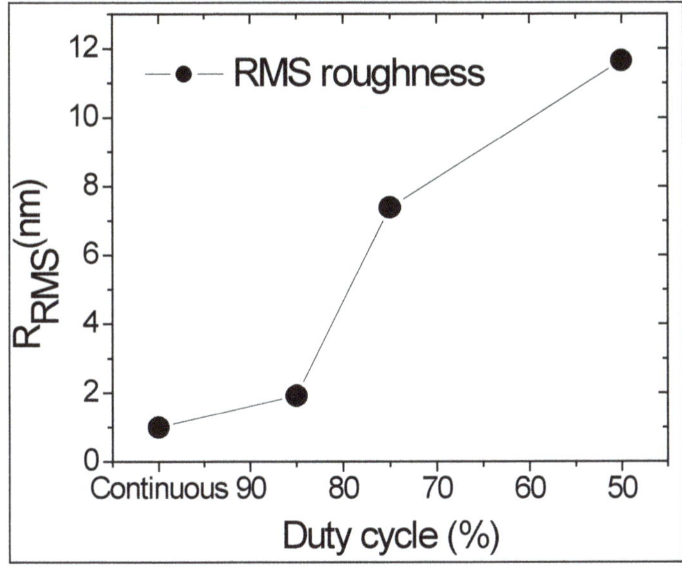

Figure 4.4: The RMS Surface Roughness of the a-SiGe:H Films Deposited with Various Duty Cycles.

4. Discussion

From the structural and optoelectronic study we have observed that with the variation of duty cycle, theformation of SiGe nanocrystallites and subsequent improvement in the optoelectronic properties. Now, we would like to address the question: *What controls the optimum microstructures in the Square wave pulsed plasma CVD deposited Silicon Germanium films*? Here, the material properties are controlled by two opposing process namely i> particle formation in the plasma and ii> ion bombardment.

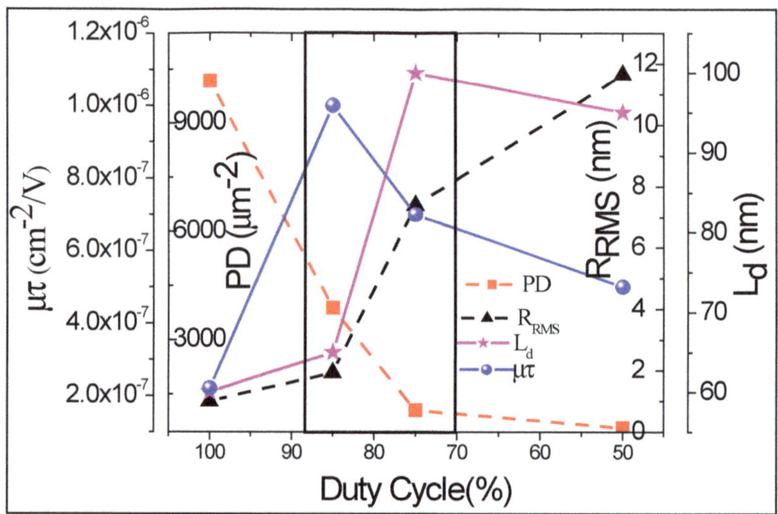

Figure 4.5: Optimization of the Film Quality in the Samples by Control of Particle Size and of the Ion Bombardment by the Variation in the Duty Cycle. The shaded region in between the two vertical dotted lines is the region for optimum deposition conditions.

With lowering of the duty cycle; particle incorporation from the plasma reduces both in number and size. On the other hand, with lowering of duty cycle (DC) the ion bombardment also decreases resulting in inhomogeneous growth. The optimum deposition condition is achieved at an intermediate value for the Duty Cycle shown by the shaded region.

5. Conclusions

In this study we have controlled the density, size and the compositions of the particles incorporated in the a-SiGe:H films deposited by rf PECVD from silane-germane mixture by playing with the plasma on time. The materials obtained at duty cycle 75 per cent is very promising with 100 nm diffusion length (initial) and 95nm (after light soaking).SWPM of the RF plasma has been found to be a very efficient technique in density, size and the compositions of the particles incorporated in the a-SiGe:H films from plasma.

6. Acknowledgements

The work has been partially funded by Indo-French Centre for the Promotion of Advanced Studies. HRTEM measurements have been carried out at the facility of the Nano Technology project at *Indian Association for the Cultivation of Science;* funded by department of science and technology, Govt. of India.

REFERENCES

1. Q. Wang and E. Iwaniczko, J. Yang, K. Lord, and S. Guha "High-Quality 10 Å/s AmorphousSilicon Germanium Alloy Solar Cells by Hot-Wire CVD" October 2001 • NREL/CP-520-30823.
2. Array D. S. Shen, J. P. Conde, V. Chu, S. Aljishi, Jeffrey Z. Liu, and Sigurd Wagner "Amorphous Silicon- Germanium Thin-Film Photodetector" IEEE Electron Device Letters, Vol. 13. No. I, January 1992.
3. Mario Moreno, Alfonso Torres, Roberto Ambrosio and Andrey Kosarev National Institute of Astrophysics, Optics and Electronics, INAOE and niversidad Autonoma de Ciudad Juarez, UACJ, Mexico "Un-Cooled Microbolometers with Amorphous Germanium-Silicon (a-GexSiy:H) Thermo-Sensing Films" Intech open.
4. K.Tanaka, E.Marayama, *et al.,* Amorphous Silicon, John Wiley and Sons, New York, (1998).
5. Sheng-Da Liu, Si-Chen Lee, and Ming-Yau Chern "Hydrogenated Amorphous Silicon-Germanium PIN X-Ray Detector" IEEE Transactions On Electron Devices, Vol. 48, No. 8, August 2001.
6. D. V. Lang, R. People, J. C. Bean and A. M. Sergent, "Measurement of the Band Gap of GexSi1–x/Si Strained Layer Heterostructures," Applied Physics Letters, Vol. 47, No. 12, 1985, pp. 1333-1335. doi:10.1063/1.96271.
7. N. Usami, T. Ichitsubo, T. Ujihara, T. Takahashi, K. Fujiwara, G. Sazaki and K. Nakajima, "Influence of the Elastic Strain on the Band Structure of Ellipsoidal SiGe Coherenty Embedded in the Si Matrix," Journal of Applied Physics, Vol. 94, No. 2, 2003, pp. 916-920. 003, pp. 916-920. doi:10.1063/1.1580194.
8. R. Braunstein, A. R. Moore and F. Herman, "Intrinsic Optical Absorption in Germanium-Silicon Alloys," Physical Review, Vol. 109, No. 3, 1958, pp. 695-710 doi:10.1103/PhysRev.109.695.
9. K. H. Jun, J. K. Rath and R. E. I. Schropp, "Enhanced Light-Absorption and Photo-Sensitivity in Amorphous Silicon Germanium/Amorphous Silicon Multilayer," Solar Energy Materials and Solar Cells, Vol. 74, No. 1-4, 2002, pp. 357-363. doi:10.1016/S0927-0248(02)00095-8.
10. J. Yang, A. Banerjee, S. Guha, "Triple-junction amorphous silicon alloy solar cell with 14.6 per cent initial and 13.0 per cent stable conversion efficiencies", Appl. Phys. Lett. 70 (1997) 2975–2977.

11. S. Guha, D. Cohen, E. Schiff, P. Stradins, P.C. Taylor, J. Yang, Industry–academia partnership helps drive commercialization of new thin-film silicon technology, Photovoltaics. Int. 13 (2011) 134–140.

12. A.H. Mahan, Y. Xu, L.M. Gedvilas, D.L. Williamson, A direct correlation between film structure and solar cell efficiency for HWCVD amorphous silicon germanium alloys, Thin Solid Films 517 (2009) 3532–3535.

13. Y. Poissant, P. Chatterjee and Pere Roca i Cabarrocas, J. Appl. Phys. **94** (2003) 7305.

14. M. E. Gueunier, J. P. Kleider, R. Bruggemann, S. Lebib, P. Roca i Cabarrocas, R. Meaudre and B. Canut, J. Appl. Phys. **92** (2002) 4959.

A. Hadjadi, A. Beorchia, P. Roca i Cabarrocas, L. Boufendi, S. Heut, and J. L. Bubendorff, J. Phys. D **34** (2001) 690.

15. Bhaduri, P. Chaudhuri, D.L Williamson, S. Vignoli, P. P. Ray and C. Longeaud, "Structural and optoelectronic properties of SiGe alloy thin films deposited by pulsed RF plasma CVD" J. Appl. Phys. **104** (2008) 063709—063709-9.

16. "A. Bhaduri, P. Chaudhuri, S. Vignoli and C. Longeaud, Correlation of structural inhomogeneities with transport properties in amorphous silicon germanium alloy thin films, Solar Energy Mat. and Sol Cells **94**(2010)1492-1495.

17. L.W. Veldhuizen, C.H.M. van der Werf, Y. Kuang, N.J. Bakker, S.J. Yun, R.E.I. Schropp, Optimization of hydrogenated amorphous silicon germanium thin films and solar cells deposited by hot wire chemical vapour deposition, Thin Solid Films, **595**(2015) 226-230.

18. Ayana Bhaduri and Partha Chaudhuri, "Fine particles in amorphous silicon germanium thin films"Phys. Status Solidi C, **7** (2010) 812-815.

Chapter 5

An Overview of Risk Management in Solar PV Projects

Subhra Das

Renewable Energy Department, Amity University, Haryana
E-mail: sdas@ggn.amity.edu

Risk Management is a central part of strategic management of any project execution. It is the process whereby project manager methodically address the risks to every aspects in terms of qualitative and quantitative risks. The solar photovoltaic projects suffer from inherent barriers of insufficient data for prudent project analysis, cost intensive structure, difficulties in guaranteeing cash flow and no enforceable securities. Risk management, if properly undertaken will increase the possibility of successful completion of project meeting the project cost, time and performance objectives.

Keywords: *Risk assessment, Risk management, Political risk, Social risk, Technical risk.*

1. Introduction

Renewable energy source is abundant and there are many promising options for converting it into useful energy. The relative merits of renewable energy vary greatly depending on the scale, capacity, and status of individual technologies, natural resource availability, characteristics, location and a number of other factors. But it is generally true that renewable energy resource is infinitely available in all regions of the world, and that the conversion efficiencies for harnessing it and costs involved have improved considerably.

Levels of investment in renewable energy projects are higher than ever. The factors driving this investment include a need to meet carbon emissions reduction targets, secure long-term energy supplies and reduce dependence on fossil fuels (Alison Riddell).

Is Risk Assessment Needed for Solar Projects?

Consider the example of Ivanpah Solar Thermal Power Plant. It is a central receiver type solar thermal power plant located in Mojave Desert at the base of Clark Mountain in California. The plant has a gross capacity of 392 MW. There are 3 units, Unit1 produces 126MW, Unit 2 and Unit 3 is of 133MW each. The construction of project started in 2010 and was commissioned in 2014. The cost of the project is $2.2 billion dollars.

Researchers at the U.S. Fish and Wildlife Service reported in Avian Mortality Report by the California Energy report, 2014 that birds while flying in the sky probably get deluded by seeing the massive reflective field of Ivanhpah solar thermal power plant and imagine it as a lake or more like thousands of puddles. The birds thus hovers into the solar field and get burnt. Other reason given by the researchers is that birds while trying to catch insects which are swarming around it gets burnt in the solar field. According to the report, 141 birds including peregrine falcon, barn owl and yellow rumped warbler were collected in October 2013 and 47 of the death were attributed to solar flux.

Figure 5.1: Ivanpah Solar Thermal Power Plant.

In April 2015, biologists estimated that 3500 birds died at Ivanpah in the span of 1 year. In September 2016, they said that 6000 birds died from collision or immolation annually. The death of hundred thousands of birds has become a concern for the wildlife officials. They consider the concentrated solar power plant as a "mega-trap" that decimates the ecosystem, first attracting insects, and then attracting birds that eat insects (S Anthony, 2014).

The Wall Street Journal on 16th March 2016 reported, "Could California's massive Ivanpah solar power plant be forced to go dark?" Environmentalist argued to decommission the power plant which cost $2.2 million dollars as it is not producing the electricity as per contract but is the cause of death of thousands of

Figure 5.2: A burned MacGillivray's Warbler that was Found at the Ivanpah Solar Plant during a Visit by U.S. Fish and Wildlife Service in October 2013.

birds. However, the solar power developers are asking the California Public Utilities Commission for permission to run Ivanpah another year to sort out problems associated with environment and technology.

The above solar thermal power plant illustrates the need of Risk Analysis and Management for Solar Projects and shows how technological or environmental risks can be a potential threat for a power plant.

2. Risk Analysis and Management

Risk Analysis and Management has their origins in the insurance industry in the USA in 1940. Project risk analysis and management is a process which enables the analysis and management of risks associated with a project. Risks and uncertainties are associated with all projects. A risk arises when it is possible to make a statistical assessment of probability of occurrence of a particular event. The actual risk can be quantified as follows:

Risk = Probability of event x Magnitude of loss/gain

Uncertainty, on the other hand, is used to describe situations when probability of occurrence of an event is not known. Thus, risk can be insured but uncertainty is not insurable. All uncertainty produces an exposure to risk which may lead to failure to: keep within the budget, achieve the required completion date and achieve the required performance objective.

Risk analysis helps decision makers to make proper mitigation plan to reduce or remove the impact of the risk. There are various types of risks associated with solar PV projects: Political Risk, Technological risk, Environmental risk, Social Risk,

Financial/Economical risk, Operational risk, Market risk. Most of these risks have their impact on the operational phase and some have impact on construction and abandon phase of the project.

3. Risk Management and Mitigation

Risk management process involves the following steps: Project definition and requirement, Risk identifications, Risk evaluation, Risk control, Risk follow-up and Risk feedback.

The risks identified during risk assessment are further characterized in terms of probability of occurrence and its impact on CAPEX, OPEX, revenues, scheduling *etc.* For most business decisions there are four main categories of risk: high probability-high impact, low probability-high impact, high probability-low impact, low probability-low impact. Waterfall diagrams are used to incorporate risk mitigation activities in the standard project management procedures. The assigned objective of a risk mitigation activity is to reduce the impact or likelihood of a specific risk factor. If a risk is high, it is unacceptable to the project, its mitigation is critical to project success. It must therefore be closely monitored by project management. A risk mitigation activity may thus be on the project's critical path, making the activity especially important.

There are four different strategies accepted in risk control which involves Avoid, Mitigate, Transfer/share and Accept. The first three strategies are proactive while the last is passive. Each control action is fed into the risk model to assess the specific impact on the business model. A risk follow up is conducted to update the changes in project over the period of time, monitor the risk control action plan and prepare a detailed report for investment stakeholders.

The choice of approach for dealing with risk analysis depends on the size, type and general nature of the project or the problem being modelled, the amount and reliability of information available and the nature of output required. The nature of the output needed will depend on the type of decisions to be made and on the particular needs of the client.

Experienced practitioners of risk analysis are quick to point out that the technique of analysis. Just as there is no such thing as a software-only solution to the problem of dealing rigorously with risk and uncertainty, so the technique used for analysis must be viewed in the context of an overall attitude of mind, which is the framework for risk management.

4. Risk Management Associated with Various Types of Risks

4.1 Political Risks

Political or regulatory risk are risks associated to a change in public policy, for example subsidies policy affecting plant profitability, Reduction in Feed in Tariffs, Increase in VAT, Delay in transferring licensing from SPV to the operator, Lack of long term planning and visibility *etc.*

According to Solar developers for example, the important issues related to policy and regulatory aspects of Jawaharlal Nehru National Solar Mission (JNNSM) was that of clarity in guidelines, bankability of Power Purchase Agreement(PPA) with NTPC Vidyut Vyapar Nigam Ltd (NVVNL), domestic content criteria and the tariff bidding process followed the PPA issue. Phase I of National Solar Mission introduced the concept of bundling of power in order to facilitate grid connected solar power generation. The mechanism of bundling is to bundle relatively expensive solar power with power from the unallocated quota of power generated at NTPC coal based stations which is relatively cheaper and sell it to the distribution utility at weighted average price.

Under NVVNL bundling scheme, price of solar is market determined as Solar Power Developers (SPDs) are given discount on CERC determined tariff. Cost of conventional power is regulated as it is sourced from unallocated quota of NTPC power plants. A PPA for 20 years is signed between NVVNL and SPDs under the bundled scheme. The PPA for bundled scheme under JNNSM is a major payment security concern for lenders.

The lenders do not find the PPA credible and bankable as it is very common that state DISCOMS fails to pay NVVNL its dues. Risk arises from the fact that because of poor and worsening financial situation of the State DISCOMS, they are unable to pay the dues to NVVNL. And NVVNL as a trader passes on the risk of non-receipt of revenues from DISCOMS for bundled power sale to the SPDs. This increases the credit risks of the lenders. Further, the PPA is signed for 20 years and there are possibilities of DISCOMS not honoring their obligations under the PPA for 20 years in case cost of solar comes down in the next 5 years.

The Ministry of New and Renewable Energy honored the concern of the lenders and introduced an additional payment security scheme for grid connected solar projects under JNNSM. Govt. of India approved Payment Security Scheme to facilitate financial closure of projects under Phase I of JNNSM by extending Gross Budgetary Support (GBS) amounting to INR 486 crore to MNRE in event of default in payment by state DISCOMS to NVVNL.

All countries across the globe are promoting the use of renewable energy for the sustainable development of the country. Various policies and regulations are made to promote renewable energy (Renewable Energy Projects Handbook, 2010). This is done keeping in mind the impact of the policies on renewable energy projects, and measures to reduce its impact. Figure 5.3 shows the Policies decided by European Union to promote Renewables (Renewable Energy Projects Handbook, 2010).

4.2 Environmental Risks

Environmental Risk is the risk of damage to the environment caused by power plant, and the liability arising from such damage. The possible impact of solar PV power plants on environment are displacement of nationally or globally threated species of animals or birds, loss of habitat for resident bird species or wildlife caused by construction operation and maintenance activities of CSP and PV, disturbance of resident bird species or wildlife caused by construction, operation and maintenance

World Energy Council — *Renewable Energy Projects Handbook*

POLICIES FOR PROMOTION OF RENEWABLES

Tables 16-20

Organisation/Country	Laws/Regulations	Promotion Programmes	Tax Arrangements/ Incentives	Environmental Policies
European Union (EU)	• Directive 2001/77/EC on the promotion of electricity produced from renewable energy sources 22.1% of electricity from REN sources in 2010 • Directive on Energy Savings in Buildings (proposal) COM (2001) 226	• White Paper for a Community Strategy and Action Plan [COM (97) 599 final (29/11/1997)] – 1,000,000 PV systems – 10,000 MW of large wind farms – 10,000 MW_{th} of biomass installations – integration of RENs in 100 communities • Green Paper: Towards a Europe Strategy for the Security of Energy Supply [COM (2001) 769 final (29/11/2000)] • "Intelligent Energy - Europe" (EIE) program (2003-2006) (came into force on August 4, 2003) : Promotion of schemes - – SAVE – energy efficiency and demand management – ALTENER – new and renewable energy sources and diversification of energy production – STEER – energy aspects of transport	Indicative financial framework for EIE €200 m (69.8 m for SAVE, 80 m for ALTENER, 32.6 m for STEER and 17.6 m for COOPENER). Contributions from expected EU enlargement is expected ~ €50 m	• European Climate Change Program (ECCP) – European strategy to implement the Kyoto Protocol. – To cut emissions by some 122-178 million tons of CO_2 equivalent – Renewable certificates trade support

Figure 5.3: Policies for Promotion of Renewables (Renewable Energy Projects Handbook, 2010).

activities and electrocution and collision caused when perching on, or flying into, the power line infrastructure.

The possible risk management stategies adapted for mitigation of environmental risks are like avoid legally protected areas and other ecologically sensitive sites such as Important Bird Areas (IBAs) and some freshwater aquatic features, landscape features such as hedgerows and mature trees should not be removed to accommodate panels and or avoid shading. If removal of a section of hedge is essential, any loss of hedges should be mitigated elsewhere on the site. All overhead power lines, wires and supports should be designed to minimize electrocution and collision risk (for example, bird deflectors may be necessary). Power lines passing through areas where there are species vulnerable to collision or electrocution should be undergrounded. Time of construction and maintenance to avoid sensitive periods (*e.g.* during the breeding season). Solar farms generally do not have moving parts, any risk to grazing animals or wildlife from moving parts that are present must be avoided.

Recognising the potential of environmental risks to affect the financial objective of energy projects, researchers and scientists have come up with innovative solutions to help solar project developers to identify site with potential environmental risks. For example, the Birdlife UNDP/GEF Migratory Soaring Birds (MSB) project has launched an innovative web tool that helps energy project developers to identify potential conflict in site selection with migratory soaring birds within the Rift

Valley or Red Sea flyway. The web tool generates known data sources of migratory soaring birds within 20km of the site, provides data based on recored satellite tracks of migratory birds, predict sensitive species that are expected to be present in the selected area and also the probability of collision of the birds.

4.3 Operations and Maintenance Risks

Operational risks are risk of unplanned plant closure, for example owing to unavailability of resources, plant damage or component failure. Table 5.1 depicts the components of Operations and Maintenance that needs to be considered during project management and their impact on project objectives.

Table 5.1: Operations and Maintenance Risks and its Impact

Risks	Components	Impact
Operations and Maintenance Risks	What are the component failure/reliability risks?	Serial failure
	Is there adequate availability of spare parts in inventory or rapidly available?	High rates of degradation
		Forced outages
	What is the strength of system design and production engineering?	Planned and unplanned maintenance downtime and costs
	Are there equipment warranties?	
	Are the manufactures still able to service them?	Manufacturer insolvency
	How strong is the O and M provider?	Resource inadequacy

4.4 Technical Risks

Technical risks are risks that are related to project designing, component selection, project feasibility study, site selection *etc.* that affects the performance of the system. Table 5.2 shows different types of technological risks associated with solar PV projetcs, the factors which needs to be studied during desinging or component selection and impact of each of these on project performance.

Table 5.2: Technical Risks and its Impact

Risk	Considerations	Impact
Resource Estimation	What level of confidence should be applied to historical solar date?	Resource related production shortfalls
		Debt service delinquency or default
Component Specifications	What is the performance history or specification of the selected product?	Manufacturer insolvency
		Serial defects
		Infant mortality
System design	How well is the system design integrated with the components?	Component failures
		Production shortfalls
	Does it ensure reliability, availability and maintainability (RAM)?	Forced downtime

Risk	Considerations	Impact
Performance Estimates and Acceptance/ Commissioning Testing	How well validated are the performance estimates? What tests are done to confirm baseline performance?	Production shortfalls relative to estimates – could stress debt service
Site Characterization	What is known and what might not be known about the site? What are the weather, water, geotechnical and infrastructural conditions?	Environmental constraints Infrastructure constraints Transmission cost overruns
Transport/ Installation risks	Are components shipped and installed according to best practices?	Equipment damage delays

Project management requires that appropriate plan should be made to address each of the risks. Some of the risk management techniques that can be adapted to reduce the impact of technical risks are proper selection of solar radiation data is required to estimate solar radiation. It has been reported (Marie Schnitzer, 2014) that modeled data sources have measurement uncertainties in the 8 to 15 per cent range. On the other hand, on site measured data has much lower measurement uncertainty, depending on the quality of instrumentation and the frequency of onsite maintenance. It is recommended that a regular schedule for leveling and cleaning solar sensors is necessary to achieve the lowest measurement uncertainty. Proper selection of Site and selection of Tier I or Tier II Manufactures and EPCs for O and M will also help in reduction of uncertainty associated with solar projects. Manufacturers may opt for Purchase warranty insurance to improve their perceived ability to be financed. Property Risk Insurance under Operational Phase, which, in addition to fire, theft and vandalism, also could potentially cover animal bites is one of the safest way of reducing impact of risks on solar projects.

4.5 Social Risks and Mitigation

Social risks are risk that affects the viability of the business due to social conflicts. Some of the social risks are:

- ✰ Modules theft leads to insurance and security cost increase
- ✰ Opposition from local community in building up power plant

It has been observed that thefts in solar PV projects happens during the construction time as well as after installation. PV modules are the main target of the theives in remote areas. It has been reported (Rebekah Hen *et al.*, 2017) that PV systems theft is increasing about 16 per cent every year which is derived from the number of insurance claims by developers. The actual number maybe more than what is being predicted since not all PV modules are insured.

The theft of solar panels can be prevented by putting fence around the solar PV systems and other basic security measures like security alarms and cameras. Also installer should follow installation recommendations while installing solar panels. The panels should be mounted on the sturcture and properly screwed to the frame instead of clipping them at the edges. Industry also provides solutions to prevent theft. Heliotex (Beck Ireland, 2010) develops customized stainless steel nuts and bolts to lock panels. The fasteners fits different types of solar rack assemblies and has a unique key that fits the head design. The key is given to the installer or owner to unfasten the hardware. Insurance of the solar panels is also recommended to reduce the impact of theft on solar developers.

In order to increase the accepantce of the project by the local community, solar developers often provide some monetary or infrastructrul benefits to the hosting town or area. This helps in building support for the facility but is not always helpful in resolving all problems. Thus, during the initial phase of site selection, efforts should be made to know the perspective of the local community towards the project and counsel them as needed. Young men or women from the nearby areas can be given employment in the energy projects as per their skills to increase acceptance of the project. So that the locals views the project as a potential employment option.

In order to strengthen project sustainability, public consultation and participation are implemented allowing for engagement of civil society both locally and internationally. Various projects are undertaken to promote solar power generation and increase its accepatnce by local community like the project by Conservation Group Royal Society for Protection of Birds and alternative energy firm Anesco. The project aims to create and restore natural habitats at solar farm sites in the U.K.

5. Conclusions

With Government promoting Solar PV projects, this sector has attracted investment from a large number of power companies. Thus, effective risk management is critical to ensure that project developers finds it beneficial to invest in these projects. Solar power developers may focus in reducing general business risk by sharing risk with joint venture partners or by investing in late stage of developments. To reduce technical risks in solar projects, developers can pool maintenance equipment and spare parts from Tier I and Tier II manufacturers as well as collect relevant weather data from reliable sources. Tie up with industry experts may enable development of epertise both within the renewable energy sector and among external stake holders and reduce risk in renewable energy projects.

REFERENCES

1. Alison Riddell, Steve Ronson, Glenn Counts, Kurt Spenser.Towards Sustainable Energy: The current Fossil Fuel problem and the prospects of Geothermal and Nuclear Power.
2. https://web.stanford.edu/class/e297c/trade_environment/energy/hfossil.html
3. Beck Ireland, 2010. Rise in solar panel theft prompts stronger security measures. http://ecmweb.com/contractor/crime- stoppers

4. Could California's massive Ivanpah solar power plant be forced to go dark? The wall Street Journal, 2016. http://www.marketwatch.com/story/could-californias-massive-ivanpah-solar-power-plant-be-forced-to-go-dark-2016-03-16
5. Financial Risk Management Instruments for Renewable Energy Projects. Summary Documents. UNEP, Division of Technology, Industry and Economics
6. Home (/en) > News (/en/news) > New web tool to integrate migratory bird conservation in development projects gets lift-o
7. https://en.wikipedia.org/wiki/Ivanpah_Solar_Power_Facility
8. Marie Schnitzer *et al.*, Reducing Uncertainty in Solar energy estimates. AWS Truepower, 07/03/2012
9. Rafter J. Risk Analysis in Project Management. Spon Press, Taylor and Francis Group, Lonon and New York
10. Rebekah Hren, Brian Mehalic, 2017. Security and Theft Prevention. http://solarprofessional.com/articles/operations-maintenance/security-and-theft-prevention
11. Report on barriers for solar power development in India. South Asia Energy Unit, Sustainable Development Department, The World Bank
12. Sebastian Anthony on August 20, 2014 at 9:35 am. California's new solar power plant is actually a death ray that's incinerating birds mid-flight. (https://www.extremetech.com/author/santhony)
13. The California Energy report, 2014 (http://docketpublic.energy.ca.gov/PublicDocuments/09-AFC-07 C/TN202538_201 40623T1 54647 _Ex h_31 07 _Kagan_et_al_201 4.pdf)
14. The Owner's Role in Project Risk Management. The National Academic Press, Washington D. C.
15. These Solar Farms Help–Not Harm–Birds and Bees. http://www.takepart.com/article/2016/03/09/solar-farm-wildlife-habitat-threatened-species
16. Travis L. DeVaulta, Thomas W. Seamansa, Jason A. Schmidt. *et al.*, Bird use of solar photovoltaic installations at US airports: Implications for aviation safety. Landscape and Urban Planning, 2014, 122, 122– 128.

Chapter 6

Evaluation on Operation of Independent Power Producer (IPP) 5 MW_p Photovoltaic Power Generator in Kupang, East Nusa Tenggara, Indonesia

Arya Rezavidi[1]*, Hamzah Hilal*[2] *and*
Petrus Tri Bakti Nurhayadi[3]

[1,2]*Agency for the Assessment and Application of Technology, Indonesia*
[3]*PT Len Industri (Persero), Indonesia*
[1]*E-mail: arya.rezavidi@bppt.go.id;* [2]*E-mail: hamzah.hilal@bppt.go.id*
[3]*Email: tribakti670@yahoo.com*

5 MWp Photovoltaic Power Plant in Kupang, East Nusa Tenggara is the first and the largest Photovoltaic Power Generator operated by Independent Power Producer (IPP) in Indonesia to date. The decision to designate PT Len Industri (Persero) as an IPP was granted by the Ministry of Energy and Mineral Resources (MEMR) of Indonesia on January 2014. The energy produce by this IPP is delivered to PT PLN (Persero), the only state owned electricity company in Indonesia. However, the Power Purchase Agreement (PPA) between PT Len Industri and PT PLN (Persero) was signed on January 9th, 2015. The electricity price in this contract is US $0.25/kWh for20 years contract period. Commercial on Date (COD) of this system was started on March 1st 2016.

The power plant is located in DesaOelpuah,KupangRegency,EastNusa Tenggara Province. Highest sun radiation measured in this area is 6.38 kWh/m^2/daywith the lowestat5.52 kWh/m^2/day. Summer season is approximately8months/year. The area covered by this plant is more than 75,000 square meters. The electrification ratio in this area is 58.91 per cent, which is thesecond lowest levelin Indonesia. About 22,008 of PV polycrystalline modules, manufactured by PT Len Industri (Persero), used for this power plant which is configured in 917 parallel strings with 24 modules in series each. Number of string inverter used in this power plant is 250 with nominal unit power of 20,000 kW. The inverter is produced by SMA, Germany.This paper is intended to publish the evaluation of operation of thePhotovoltaic Power Generator, after one year operation. During seven months observation, energy produce by this power plant is far below the target. During the full system operation the average daily energy measured is 22,416 kWh with maximum energy measured of 26,253 kWh.

Keywords: *Photovoltaic, Electrification ratio, IPP, Kupang, East Nusa Tenggara, Indonesia.*

1. Introduction

Indonesia is the largest archipelago country in the world with population reaches 250 million. Numbers of island in Indonesia is about 13,500, and about 60 per cent out of it is inhabitant. Only five islands are considered as large islands. Among the large islands, only in Java that the village's settlement are quite dense and close together. Sixty per cents of the population of Indonesia lives onthe island of Java, even though the land area of Java counts only 20 per cent of the total land area of Indonesia. In the other major islands the settlement are scatteredand often isolated far from the nearest town. With such a condition, to electrify the entire populated islandsis not an easy task. Until today there are still about 11 per cent of the population in Indonesia do not yet have access to the electricity networks provided by the Government.

In the mid2016 Indonesia average electrification ratio is nearly 89.5 per cent which means there are still about 7 million of households do not have access to electricity. If it is multiplied by average 4 persons for each family, it means almost 28 million peoplesdo not have access to the electricity. Through some electrical acceleration programs, the Government of Indonesia has been trying to increase the electrification ratio. Among the programs is to encourage the utilization of solar power generator in rural area and isolated islands. In the early 1990s the Government introduced individual solar power generator or known as Solar Home System to the community. Eventually the program was not going well due to scarcity ofspare parts supply, especially battery,and lack of technician to support maintenancein remote and isolated area. In late 1990's Government began to introduce centralized solar power plant which is combines with other generator such as Diesel generator, known as Photovoltaic Hybrid Generator. However, most of the projects are Government or PT PLN,the only state owned electricity company in Indonesia, funded programs.

In early 2012 the Government intoduced private sector's participation in operating Photovoltaic solar power generator or known as the Photovoltaic Independent Power Producer (IPP). Electricity generated by thisIPP is then purchased by PT PLN, and the energy price purchased is set through the government regulation or known as Feed in Tariff policy. At that year there are ten locations were offered to the private companies, nonetheless, only 2 IPP'sare end up with contract in 2 different locations.First is in Gorontalo, Sulawesi with 2 MW of capacity and Kupang with a capacity of 5 MW.

Independent Power Producer (IPP) 5 MWp Photovoltaic Power Generation in Kupang, East Nusa Tenggara yet becomes the first and largest Photovoltaic Power Generation operated in Indonesia to date. The decision to designate PT Len Industri (Persero) as the winner was granted by the Ministry of Energy and Mineral Resources

(MEMR) of Indonesia on January 2014 after selecting among several bidders for this project. The energy produce by this IPP is delivered to PT PLN (Persero), the only state owned electricity company in Indonesia. However, the Power Purchase Agreement (PPA) between PT Len Industri and PT PLN (Persero) was signed one year later on January 9th 2015. The electricity price in this contract is US $0.25/kWh for20 years contract period. The PPA Agreement was signed by General Manager PT PLN of East Nusa Tenggara Area Richard Safkaur and President Director of PT Len Industri (Persero), Abraham Mose in the headquarter of PT PLN in Jakarta. Commercial on Date (COD) of this system was started on March 1st 2016.

2. Electricity Conditions in Kupang, NTT

The power plant is located in Oelpuah Village, Kupang Regency, East Nusa Tenggara Province, which has geographical coordinate of 123°44′55″ East and 10°8′58″ South. Kupang Regency consists of 24 districts, and has land area of 5.431.23 km or 543,123 Ha. Geographically, Kupang Regency is located between 121°15′-124°11′ East Longitude and between 9°19′-10°57′ South Latitude. The Regency of Kupang generally has tropical climate and dry, and likely to be affected by wind known as semi arid area where rain fall is relatively low. Vegetation is predominantly savannas and steppes. Rainy season is very short *i.e.* 3-5 months per year, while the drought season almost 7-8 months. The rainy season usually occurs during the month of December until March. Atmosphere pressure ranged between 926.3 millibars, the speed of wind average reaches 6 knots/hour. The temperature of the air ranges from 24 –32 C with an average humidity of air 75 per cent-76 per cent RH. Highest sun radiation measured in this area is 6.38 kWh/m²/day with the lowest at 5.52 kWh/m²/day. Topography of Kupang Regency is mountain and hilly area with degrees of slope about 45 . The ground is so critical and sensitive to erosion. However, in low land area which is relatively fertile, people usually live

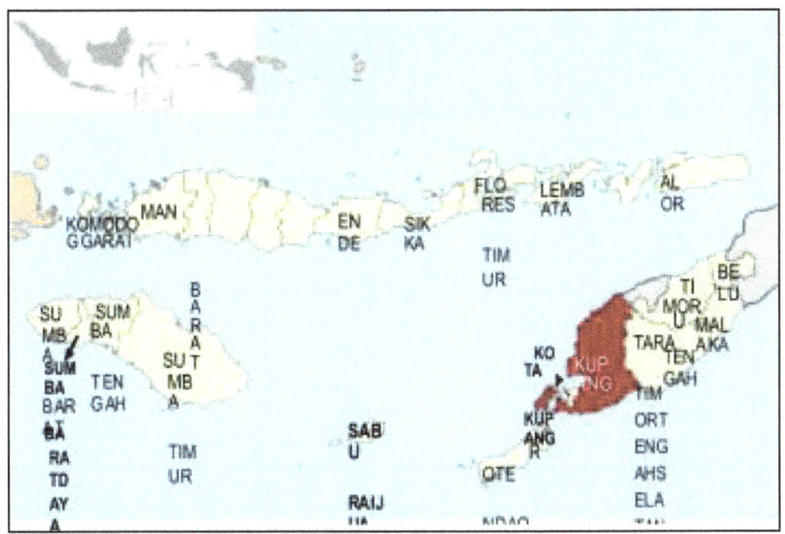

Figure 6.1: Map of East Nusa Tenggara Province.

Figure 6.2: Area of Kupang Regency.

and concentrated there. Topography like this contributes to physical, economic and social isolation, more over lack of support infrastructure such as roads and bridges in various districts. The use of inter island transportation are considered expensive because of the low frequencies of transportation between the Islands, which surely also affect the price of goods and services in particular islands. Below is a map of the province of NTT and Kupang where the Photovoltaic generator is located.

The electrification ratio in this area is 58.91 per cent, which is thesecond lowest levelin Indonesia. This area is served by PT PLN (Persero), the state owned the electricity company, Kupang Area as one unit of PT PLN (Persero) Region of East Nusa Tenggara (NTT). Table 6.1 shows the number of customers in the region of

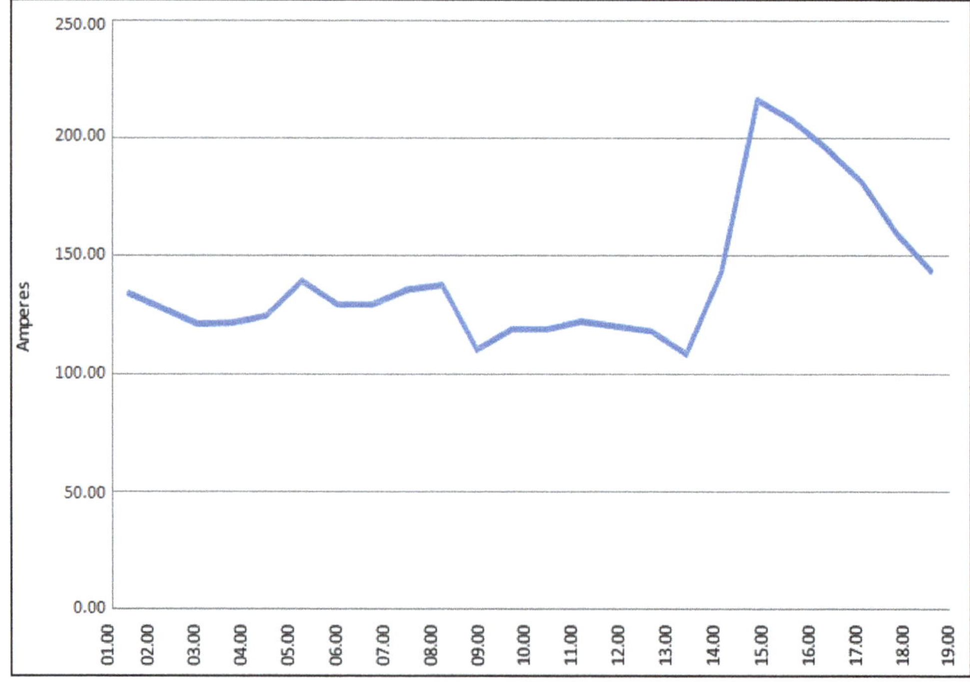

Figure 6.3: Average Daily Load Curve at Kupang.

Table 6.1: Number of Customers in the Region of PLN Kupang Area in 2013

Sl.No.	RAYON	YEAR 2013											
		JAN	FEB	MAR	APR	MAY	JUN	JUL	AGT	SEP	OCT	NOV	DES
1	KUPANG	79,967	80,087	80,310	80,916	82,391	83,562	84,444	85,700	86,459	87,273	88,286	89,564
2	OESAO	35,584	35,831	35,998	36,287	36,532	36,621	36,651	36,727	36,942	37,025	37,121	37,301

PLN Kupang area in 2013, before the Photovoltaic power generator was introduced, where the largest number of electricity use is household. Customer's growth of in the area of Rayon Oesao is rather slow and low. Figure 6.3 shows the daily load curve measured at 20 kV point in Maulafa Main Traffo before the Photovoltaic power generator was introduced.

At the time initial electrification study conducted in Kupang prior to Photovoltaic power generator was introduced, the electricity networks in Kupang is supplied mostly by Diesel Generators capable of generating to maximum power of 58,715 MW. Peak loads during the day was 45,088 MW and standby power during the day was maximum 13, 627MW. Most of the Diesel Generators are operated manually.

3. Configruation of Thephotovoltaic Generator

The area covered by this plant is more than 75,000 square meters.About 22,008 of PV polycrystalline modules, manufactured by PT Len Industri (Persero), used for this power plant which is configured in 917 parallel strings with 24 modules in series each. Number of string inverter used in this power plant is 250 with nominal unit power of 20,000 kW.

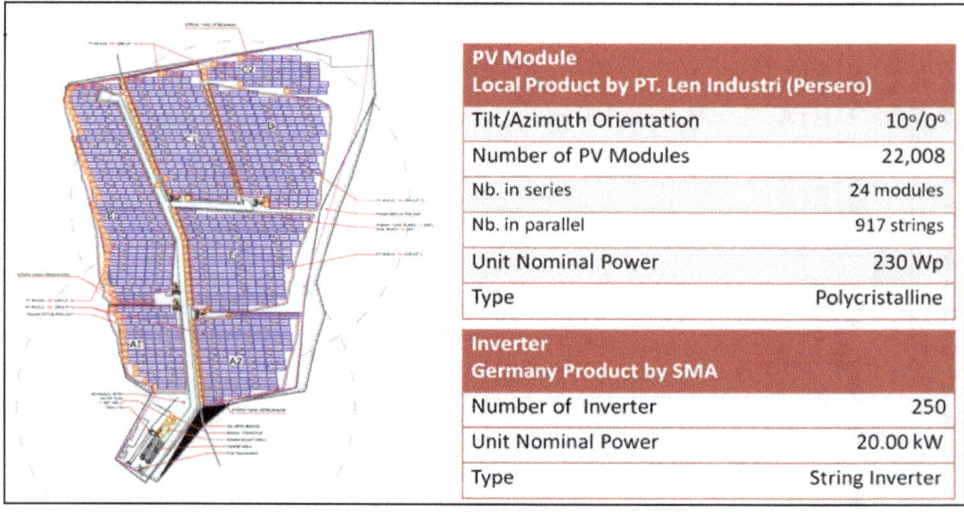

PV Module Local Product by PT. Len Industri (Persero)	
Tilt/Azimuth Orientation	10°/0°
Number of PV Modules	22,008
Nb. in series	24 modules
Nb. in parallel	917 strings
Unit Nominal Power	230 Wp
Type	Polycristalline

Inverter Germany Product by SMA	
Number of Inverter	250
Unit Nominal Power	20.00 kW
Type	String Inverter

Figure 6.4: Site Layout of Photovoltaic Generator.

The Photovoltaic system is divided into 5 blocks with capacity of 1 MW each, and uses an inverter type string with has capacity 20 kW each.

The inverter used is 20000 STP TL-30 type produced by SMA. Photovoltaic module specifications are as follows:

- ✰ Solar panel used is 230P type manufactured by PT Len Industri
- ✰ Solar cell type is the polycrystalline
- ✰ Solar panels rigs does not use foundation and designed to be mounted on a slope contours

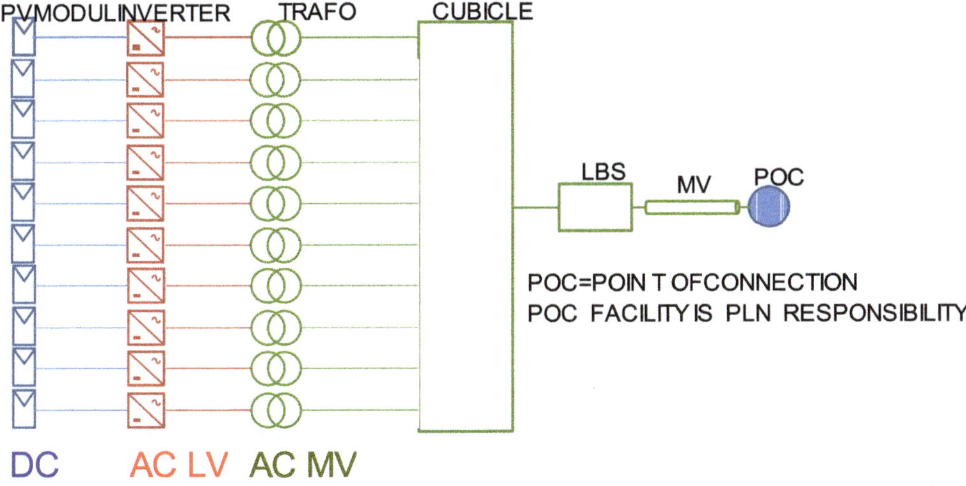

Figure 6.5: Block Diagram of Photovoltaic Generator.

The inverter is manufactured by SMA, Germany. The Photovoltaic generator is connected to the medium voltage grid (20 kV) belongs to PT PLN. Before the network points of connection there are some equipment installed, namely:

- ✰ Step up transformer to the medium voltage
- ✰ Protection panel (cubicle) and medium voltage isolator (LBS)
- ✰ Medium voltage grid to connect to the point of connection of PT PLN grid

Table 6.2: Detail Technical Specification of the Photovoltaic Generator

Capacity	:	5 MWp
Interconnection Voltage	:	20 kV
Interconnection point to point distance	:	± 5,8 km
Solar module:		
✰ Type	:	***Polycrystalline***
✰ Module capacity	:	250 Wp
Inverter		
✰ Type	:	Sunny Tripower 20000 TL
✰ Capacity	:	20 kW
✰ Efficiency	:	98,2 per cent
✰ Range of voltage	:	160-280 VAC
✰ Range of frequency	:	44 -55 Hz
Transformer		
✰ Type	:	***Indoor-Cast Resin***
✰ Capacity	:	10x630 kVA

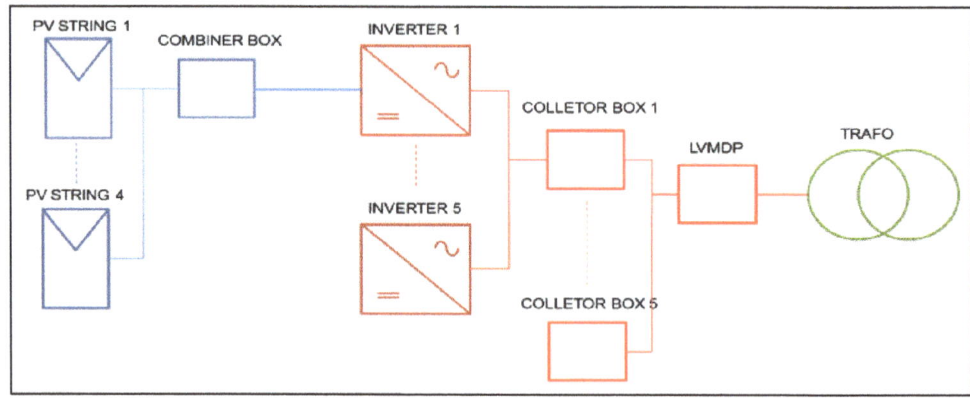

Figure 6.6: Detail Block Diagram of DC and Low Voltage AC Sides.

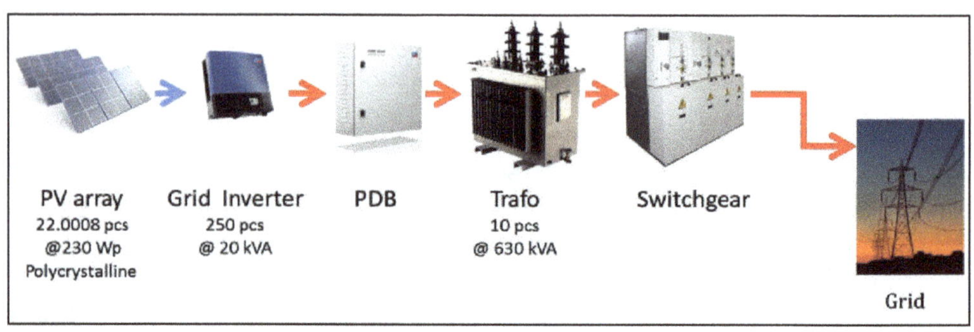

Figure 6.7: Power Plant Configuration.

Figure 6.8: Eelectricity Grid from the Power Plant to the Nearest Medium Voltage grid of PLN.

Figure 6.9: Aerial Picture of Photovoltaicarray in Kupang.

Figure 6.8 is the map of Power Plant's location seen from Google Earth. The green line is the electricity grid from the Power Plant to the nearest PLN medium voltage grid.

4. Evaluaton of One Year Operation of the PV System

This paper is mainly intended to publish the evaluation of operation of 5 MW Kupang Photovoltaic Power Generator, the largest PV IPP in Indonesia to date, after one year operation. Calculated data shows that from the average sun radiation of 2,323 kWh/m^2, the expected energy harvested for one year is about 9,133,959 kWh or 9,134 MWh. Energy losses is due to irradiance level (3 per cent), temperature (10.7 per cent), array soiling (3.2 per cent), module array mismatch (2,2 per cent), ohmic wiring (1.2 per cent), inverter loss during operation (2.3 per cent), etcetera.

During seven months observation, energy produce by this power plant is far below the target. During the full system operation the average daily energy measured is 22,416 kWh with maximum energy measured of 26,253 kWh. The highest power output measured from the generator is 4.6 MWp (measured) and it was occured on December 19, 2016.

During the first year of Photovolatic system operation, the generator never operates at maximum conditions. The average power generated only ranges slightly above 2 MWp. Restriction comes from dispatchers of PT PLN. In some cases the frequency of the grid was fluctuated corresponding to the fluctuation of power from the PV generator. In the other occasion connection fails caused by fluctuation of frequencyof the generator, so it reached the UFR setting.

There is a suspicion that there is a deviation between actual condition and the condition of the system during the interconnection study. Cloudy weather condition also often occur during months of operation.

Figure 6.10: Inauguration of the Power Plant by the President of the Republic Indonesia.

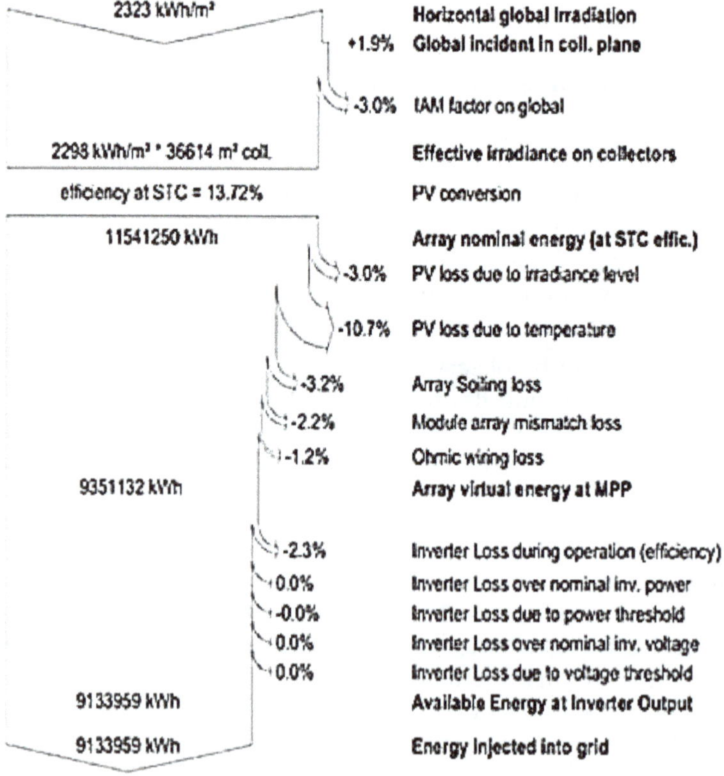

Figure 6.11: Efficiency of the System.

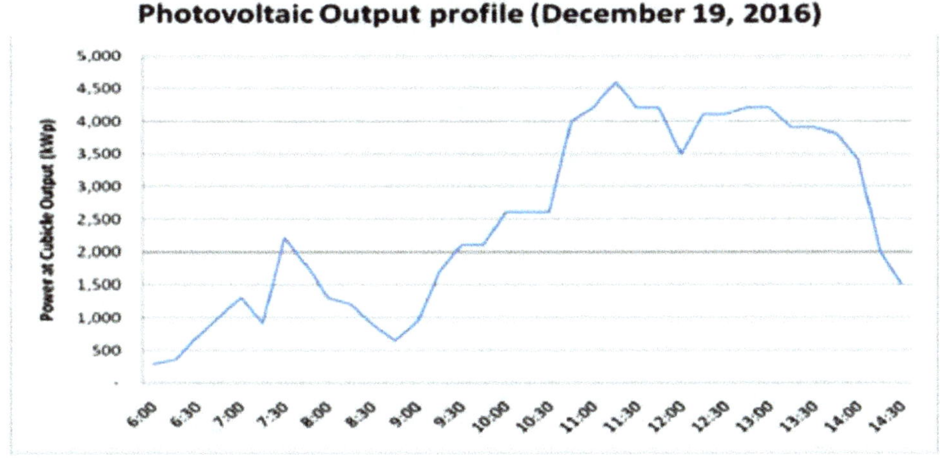

Figure 6.12: The Highest Power Output Ever Measured.

5. Conclusion

The territory of East Nusa Tenggara (NTT) Province have natural energy resources from solar energy with the longest exposure in Indonesia, e.g eight months of sun exposure in a year which can be utilized as potential renewable energy resources. With the operation of the 5 MWp IPP Photovoltaic power generator in Kupang, as many as 5,550 houses can be electrified. The existence of the IPP Project belongs to PT Len industries (Persero) in Kupang, NTT, supports the Government's

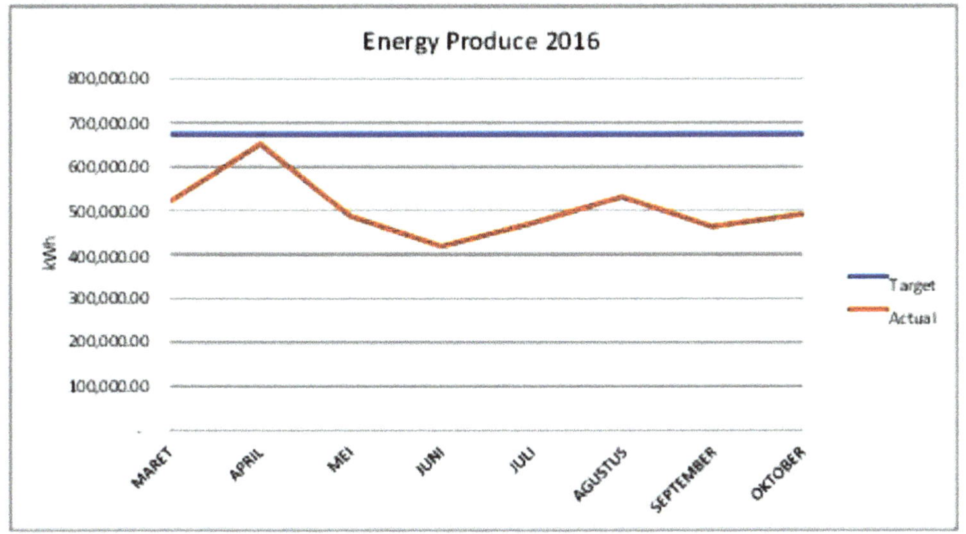

Figure 6.13: Energy Produce Always Below the Energy Targeted.

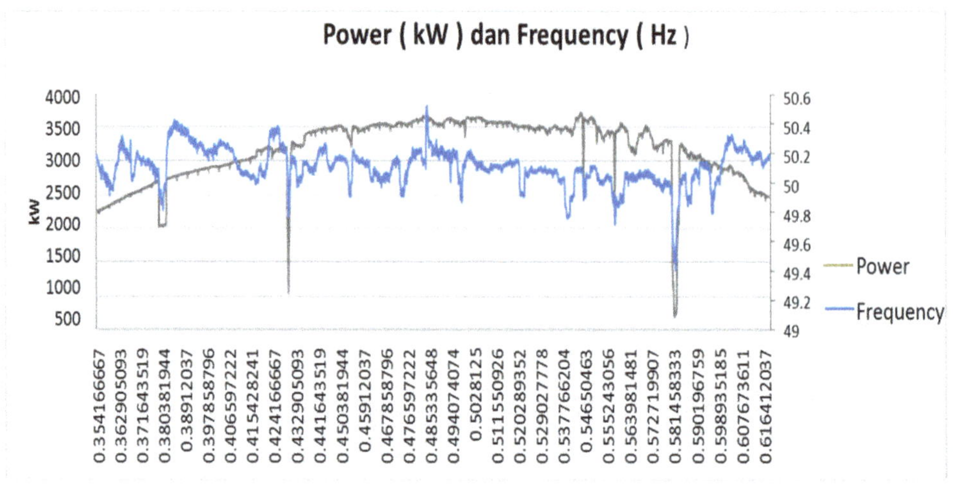

Figure 6.14: Full System Operation on November 20th 2016.

policy in the field of economic and national development in general, and in particular in the field of electronic industry and infrastructure.

Generally there are several reasons that the energy target production cannotbe reached, *e.g.*

- Limitation of power penetration to the grid by the dispatcher from PT PLN
- The frequency of the grid was fluctuated corresponding to the fluctuation of power from the PV generator
- System frequency reach the UFR setting so that the connection to the grid was tripped
- Cloudy weather often occur during the months of operation
- There is a suspicion that there is a deviation between actual condition and the condition of the system during the interconnection study.

Nevertheless, the private participation program on Photovoltaics power plant based IPP is currently halted, because the Ministerial decree associated with the Feed-in-Tariff policy is replaced with other policy that restricts the selling price of energy to PT PLN. This is a set back to the Renewable Energy program, because without this policy then the Government target to achieve New and Renewable Energy utilization amounted to 23 per cent by 2025 will be difficult to achieve.

6. Acknowledgments

The authors would like to give highly appreciation to:

1. Government of Indonesia through Ministry of Research, Technology and

Higher Education
2. Non Aligned Movement Science and Technology Center
3. Agency for the Assessment of Tehnology (BPPT)
4. PT Len Industri (Persero)

Without their support this report would never be happened.

REFERENCES

1. PT Surya Energi Intotama, Detail Engineering Design IPP Kupang, 2014
2. PT Prima Layanan Nasional Enjiniring, Pengadaan Jasa Studi Interkoneksi IPP PLTS 5 MW Kupang, Oktober 2014
3. PetrusTri Bakti Nurhayadi, Independent Power Producer (IPP) PLTS 5 MWp di Kupang, PT Len Industri, 2017
4. http://economy.okezone.com/read/2016/11/18/320/1544739/rasio-elektrifikasi-indonesia-baru-89-5-kalah-dari-vietnam-dan-thailand
5. http://www.djk.esdm.go.id/pdf/Buku per cent 20Statistik per cent 20Ketenagalistrikan/Statistik per cent 20Ketenagalistrikan per cent 202015.pdf

Chapter 7

A Review on Solar Thermal Technologies for Low and Medium Temperature Industrial Process Heat

Mohd Fauzi Ismail

Industrial Centre of Innovation in Energy Management,
SIRIM Industrial Research,
1, Persiaran Dato' Menteri, Section 2,
P.O. Box 7035, 40700, Shah Alam, Selangor, Malaysia
E-mail: mfauzi@sirim.my

Solar heat for industrial process has a large potential to be implemented in Malaysia especially for low and medium temperature application. This article reviews solar thermal technologies for industrial process heat in term of type of solar collector technologies for low and medium operation temperature, its potential application in selected industrial sectors and type of industrial processes. Then, it discusses the potential of this solar process heat specifically in Malaysia based on the energy demand data. It also provides information on how the solar energy can be possibly integrated into industrial process heat. Finally, the article presents four solar process heat demonstration plants that have been installed worldwide.

Keywords: *Solar heat for industrial process, Low and medium temperature, Solar collectors, Solar heat integration, Malaysia's potential.*

1. Introduction

Malaysia's National Energy Balance 2013 (Malaysia, 2013) reported that total final energy consumption for Malaysia in 2013 was 51,584 ktoe. According to sectors, 43.3 per cent of this final energy consumption was used by transport, 26.2 per cent by industry, 14.4 per cent by commercial/residential, 14.1 per cent from non-energy use and 2 per cent from agricultural sector. Focusing on industrial sector alone, the heating requirement accounts for a large portion *i.e.* 67 per cent of the total energy use from the fossil fuel, and the balance was for electricity. Looking at this figure, a significant energy from fuel can be reduced if some portions of the heating use in industry is supplied by Renewable Energy (RE).

The enablingpolicyframework and supportprogrammes in Malaysia for RE have focused on grid electricity power generation over thermal applications despite the facts that large portion of the energy is expended for meeting heating requirement especially in industrial sector. Moreover, many programmes, incentives and Research, Development and Innovation (R&D and I) activities related to solar energy in Malaysia have been emphasised mainly on solar photovoltaic (PV) for electricity generation and much less on heat application. Furthermore, apart from domestic solar water heaters, the government does not yet have policies, incentives or standards that specifically aim at larger-scale solar thermal system applications in commercial buildings or in industrial applications.

Malaysia is located in the equatorial region and has a tropical rainforest climate. The tropical climate has been categorized as having heavy rainfall, constantly high temperature and relative humidity throughout the year. Being a country that is close to equator, Malaysia naturally has abundant sunshine and thus solar irradiance. It is reported that annual average daily solar irradiations for Malaysia were from 3.73 kWh/m^2 to 5.11 kWh/m^2(Engel-Cox, 2012) with the highest usually recorded in March or April and the lowest is in November or December during monsoon season. Northern region and few places in East Malaysia have the highest average daily solar radiation. With this plenty of sunlight throughout the year, Malaysia has big potential for solar energy application.

Given the fact that 30 per cent of the total industrial process heat demand requires temperature below 100 °C (ECOHEAT, 2006), which can be met by commercially available solar thermal collectors, in principle the potential of solar thermal in industry is enormous (Mat, 2015). Furthermore, industrial sector has more than 80 per cent of untapped potential of energy efficiency in the period to 2035 as reported in the New Polices Scenario by International Energy Agency (IEA) (IEA, 2014b).

This paper reviews solar thermal technologies for industrial process heat in term of type of solar collectors, its potential application in selected industrial sectors, process, operating temperature, and possible points for solar heat integration. It also provides several examples of solar process heat plants in several countries including Malaysia.

2. Solar Thermal Collector Technologies for Process Heat

Solar collector in a solar thermal system is a component that absorbs solar irradiation as heat and transfers the heat to a working fluids such as water, air and thermo oil. The heat carried by the working fluid is then used to provide hot water, steam or space heating.

There are three major categories of solar thermal collectors *i.e.* 1) non-tracking (stationary), 2) single-axis tracking and 3) two-axis tracking collectors (Kalogirou, 2003). The major types of collector that can be used for industrial process heat are from the non-tracking category and one-axis -tracking collectors like scheffler dish, parabolic trough collectors and Fresnel collectors (Kalogirou, 2003). Three types of collectors that fall into stationary category are: 1) flat-plate collectors, 2) evacuated-tube collectors and 3) stationary compound parabolic collectors. The flate plate collectors are designed for low temperature (< 100 °C), meanwhile evacuated-tube collectors are for medium temperature (100°C–400°C) application. The non-tracking and one axis parabolic trough collectors (PTCs) are used for medium and high temperature (> 400 °C) industrial process heat. On the other hand, two-axis tracking collectors such as parabolic dish reflector (PDR) and heliostat field collector are used to produce steam for power generation. Table 7.1 shows the type of solar thermal collectors and their suitable applications.

Table 7.1: Type of Solar Collectors and its Application (Faninger, 2010; Kalogirou, 2003; Mekhilefa, 2011)

Motion	***Collector Type***	***Absorber Type***	***Concen-tration Ratio***	***Indicative Temperature Range (ºC)***	***Possible Application***
Stationary	Flat plate collector (FPC)	Flat	1	30-80	Pool heating, crop drying, low temp. industrial process heat
	Evacuated tube collector (ETC)	Flat	1	50-200	water, space heating, space cooling, med. temp. industrial process heat
	Compound parabolic collector (CPC)	Tubular	1-5	60-240	water, space heating, space cooling
Single-axis tracking			5-15	60-300	
	Fresnel lense tracking (FLC)	Tubular	10-40	60-250	high temp. industrial process heat
	Parabolic trough collector (PTC)	Tubular	15-45	60-300	high temp. industrial process heat
	Cylindrical trough collector (CTC)	Tubular	10-50	60-300	high temp. industrial process heat
Two-axis tracking	Parabolic dish collector (PDS)	Point	100-1000	100-500	power generation
	Heliostat field collector (HFC)	Point	100-1500	150-2000	power generation

Note: temp. (temperature), med. (medium)

In the following sub-section, this paper will focus on the technology of flat plate and evacuated tube collectors that are suitable to be used for industrial process heat which requires low and medium heating temperature. This is also because these type of collectors are suitable to be applied in countries such as Malaysia which has high cloud coverage and diffuse solar radiation. The potential of solar thermal for process heat in Malaysia will be discussed further in the next section of this paper.

Flat Plate Collectors

Flat plate collectors are the most commonly used collectors in Europe, which is the second largest market place for solar thermal collectors after China. It is reported that, 83.8 per cent of solar thermal installed capacity in Europe in 2013, being the flat plate collectors (Mauthner, 2015). This is mainly because the flat plate collectors are cheaper, require less maintenance and suitable for delivering thermal energy at temperatures between 30 °C to 80 °C (Cottret, 2010). Construction of a typical flat plate collector consists of glazing covers, absorber plates, insulation layers and recuperating tubes which is filled with heat transfer material such as water or water/glycol mixture (Tian, 2013) as shown in Figure 7.1. The principle of the flat plate collector is: when the solar irradiation hit the surface of the collector, the radiation will pass through the transparent cover and reach the absorber plate. The radiation is then absorbed by the plate and converted into thermal energy which is then transferred to the heat transfer material fluid within the tubes.

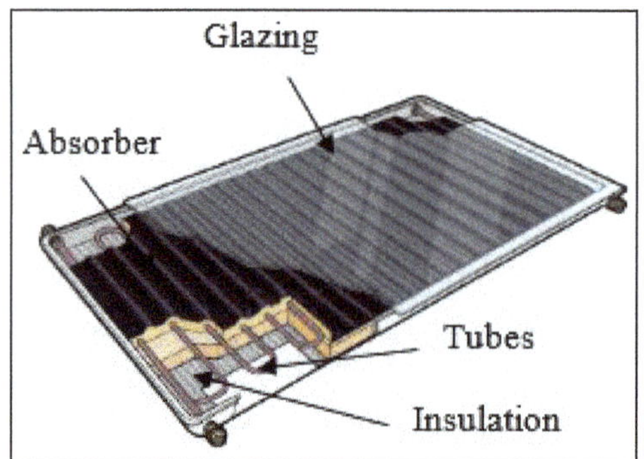

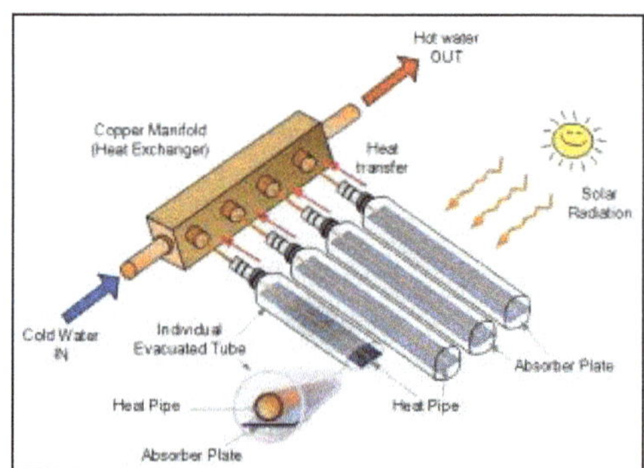

Figure 7.1: Construction of Flat Plate Collector (Left) (Trust, 2015) and Evacuated Tube Collector (Right) (Tutorials, 2015).

A standard flat plate collectors have high heat losses and not suitable for higher operation temperature (Cottret, 2010). Therefore, some efforts to improve the performance of flat plat collector in reducing thermal losses and keeping high optical efficiency were carried out. Some technological improvements include replacing the single glass by multiple and different kind of glasses and anti-reflective coating (Dagdougui, 2011; Ehrmann, 2012), modelling a gas-filled flat plate collector (Vestlund, 2009), improving structure by integrating heat pipe (Wei, 2013) and choosing better material for heat transfer (Chen, 2010).

Evacuated Tube Collectors

Evacuated tube collectors are the predominant solar thermal collector technology worldwide in 2013 with a share of 70.5 per cent from total solar thermal installed capacity (Mauthner, 2015). This figure is mainly contributed by China which is the biggest user of evacuated tube collectors, as well as the world largest installed capacity of solar thermal. Evacuated tube collectors are designed to operate at higher temperature than flat plate collector ranging from 50 °C to 130 °C (Cottret,

2010). The manufacturing process, mechanical complexity and material selection of the evacuated tube collectors are more expensive than the flat plate collectors (Trust, 2015). This is also due to the design of collectors' housing that is made of a vacuum glass tubes to reduce and eliminate convection and conduction thermal losses. Furthermore, the glass tube is used because it able to withstand the stress of the vacuum. A typical construction of evacuated tube collector is shown in Figure 7.2.

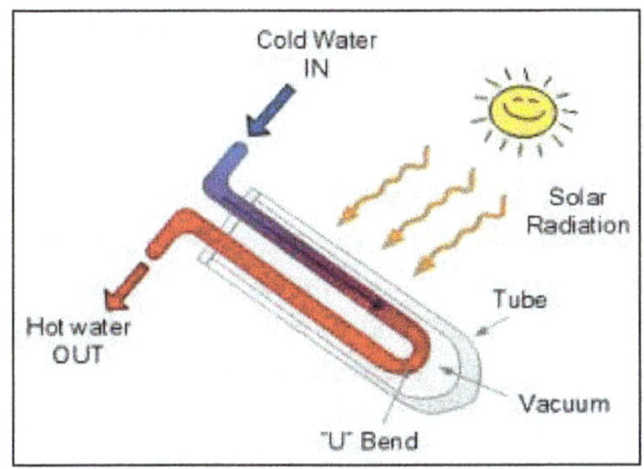

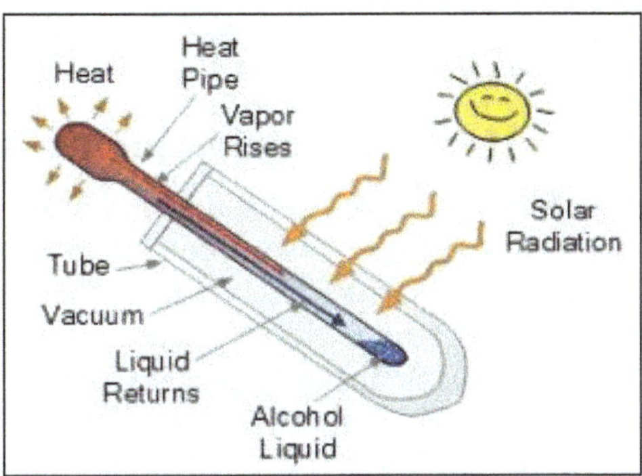

Figure 7.2: Direct Flow (Left) and Heat Pipe Evacuated Tube Collector (Right) (Tutorials, 2015).

There are two types of evacuated tube collectors *i.e.* 1) direct flow and 2) heat pipe collector. The direct flow evacuated tube collectors use an evacuated tube inside a U-shape tube. In direct flow evacuated tubes, there are two heat pipes running through the centre of the tube, one is flow pipe and the other one is the return pipe. A heat absorbing reflective plate is in between the flow and the return pipes through the solar collector tubes. The heat pipes and the reflector pipe are made out of copper with a selected coating material. Both the absorber plate and the heat transfer tube are made vacuum sealed inside a glass tube. This is shown in Figure 7.2. The heat transfer material is usually fluid and this fluid runs through concentratic tube-in-tube or a U-shaped tube to the base of the glass bulb and then returns to the header.

A heat pipe evacuated tube collectors contains alcohol or water in a vacuum which is used to absorb the energy from the sun. Due to presence of the vacuum, the alcohol or water will evaporate at a low temperature of 25 °C to form a vapour. This vapour then rises up to the top of the collector tube, heating it up and transfer the heat to the solar fluid. After the heat transfer process completed, the vapour condenses back to a liquid and flows down back to the bottom of the collector tube.

3. Solar Heat for Industrial Process

Solar heat for industrial processes is still in the infancy stage of development, but is considered to have a huge potential for solar thermal applications. It was reported that in 2006, there were 90 operating solar thermal systems for process heat worldwide, with a total capacity of about 25 MWth (35,000 m^2) and this figure accounts for only 0.02 per cent of the total solar thermal installed capacity worldwide

(IEA, 2008). The figure had increased in 2014 as 120 operating solar thermal systems for process heat have been installed worldwide which is equivalent to a total capacity of about 88 MWth (125,000 m^2) (IEA, 2014a). Nevertheless, a huge potential is still available as more than 80 per cent of untapped potential of energy efficiency is seen for the industries in the period to 2035 (IEA, 2014b). This is supported by a study (Lauterbach, 2012) that estimates the potential of solar heat in industrial process in five European countries *i.e.* Austria, Italy, Netherlands, Portugal and Spain is in the range of 3.0 per cent to 4.5 per cent of the industrial heat demand and in total of 16.7 TWh (IEA, 2008).

Solar Heat Potential in Industrial Sectors

Majority of the solar thermal application today *i.e.* 94 per cent is for domestic hot water systems (Mauthner, 2015). Although the domestic sector offers a great potential demand for solar thermal application, the industrial sector should not be left out. This is due to two important factors (IEA, 2008). First, the industrial sector covering about 26 per cent of the final energy consumption in Europe in 2012 (Agency, 2015). Second, the major share of heating energy needed in industrial sector is for low (20 °C – 100 °C) and medium temperature (~250 °C) (Schnitzer, 2007), which is a temperature that could be supplied by available solar thermal technologies (Kalogirou, 2003). A study by (ECOHEAT, 2006) reports that 30 per cent of the total industrial heat demand requires temperature below 100 °C and 60 per cent requires temperature above 100 °C. These percentages can be further breakdown into low temperature below 100 °C (30 per cent), medium temperature between 100 °C to 400 °C (27 per cent) and high temperature over 400 °C (43 per cent) (ECOHEAT, 2006) as shown in Figure 7.3. A slightly different fraction of temperature range is reported for Germany (Lauterbach, 2012) *i.e.* 21 per cent is for process heat lower than 100 °C, 8 per cent is for temperature range of 100 °C - 200 °C, a small percentage is for application between 200 °C - 300 °C and the biggest portion *i.e.* 65 per cent is needed at temperature over 500 °C.

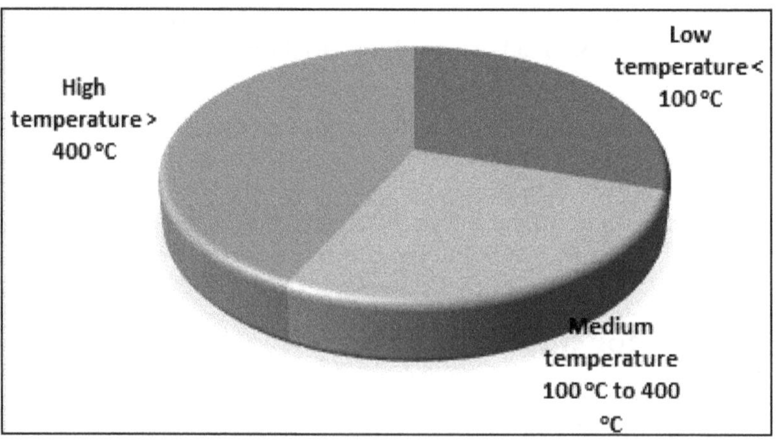

Figure 7.3: Share of Industrial Heat Demand According to Temperature Level (ECOHEAT, 2006).

A survey conducted by (IEA, 2008) has identified several potential industrial sectors and industrial processes where the solar thermal could be optimally and efficiently used. These sectors and processes are identified based on the continuous heating demand needs and the temperature level of the processes that is compatible with the operating temperature of solar thermal collectors. The sectors include food (including wine and beverage), textile, transport equipment, metal and plastic treatment and chemical. The most suitable industrial processes include cleaning, drying, evaporation and distillation, blanching, pasteurization, sterilisation, cooking, melting, painting, and surface treatment. Another study (Kalogirou, 2003) states that most of the energy used for industry in Cyprus is from the food industry and the manufacture of non-metallic mineral products. Among the food industries, milk and breweries industries have great potential to employ solar process heat as the industries involve applications such as drying, cooking, cleaning, extraction and many others. Other potential industrial sectors with its corresponding suitable process and operating temperature level (Weiss, 2015) are shown in Table 7.2.

Table 7.2: Potential Industrial Sectors with Suitable Industrial Processes and Temperature Level (Weiss, 2015)

Industrial Sector	***Process***	***Temperature Level (°C)***
Food and Beverages	Drying	30 - 90
	Washing	40 - 80
	Pasteurizing	80 - 110
	Boiling	95 - 105
	Sterilising	140 - 150
	Heat Treatment	40 - 60
Textile Industry	Washing	40 - 80
	Bleaching	60 - 100
	Dyeing	100 - 160
Chemical Industry	Boiling	95 - 105
	Distilling	110 - 300
	Various chem. Processes	120 - 180
All sectors	Pre-heating of Boiler Feed-water	30 - 100
	Heating of factory buildings	30 - 80

Solar Heat Potential Impact for Malaysia's Industry

A similar trend of final energy consumption by industrial sector in Europe was seen in Malaysia (Malaysia, 2013) *i.e.* 26.2 per cent of the total final energy demand in 2013 was used by industry. This shows that Malaysia's industrial sectors also has a great potential for solar thermal application. In contrast with statistic reported for Cyprus (Kalogirou, 2003), most of the industrial energy demand in Malaysia are from chemical and followed by food, beverages and tobacco in second place

(Malaysia, 2013) as shown in Figure 7.4. Table 7.2 indicates that low and medium temperature solar process heat are suitable for industrial sectors in Malaysia. This is supported by the fact that Malaysia's climate with significant amount of cloud and has Direct Normal Irradiance (DNI) below the required amount of concentrating collectors to be economically feasible *i.e.* 1,900 – 2,000 kWh/m²/year is not suitable for producing heat at higher temperature (Affandi, 2013).

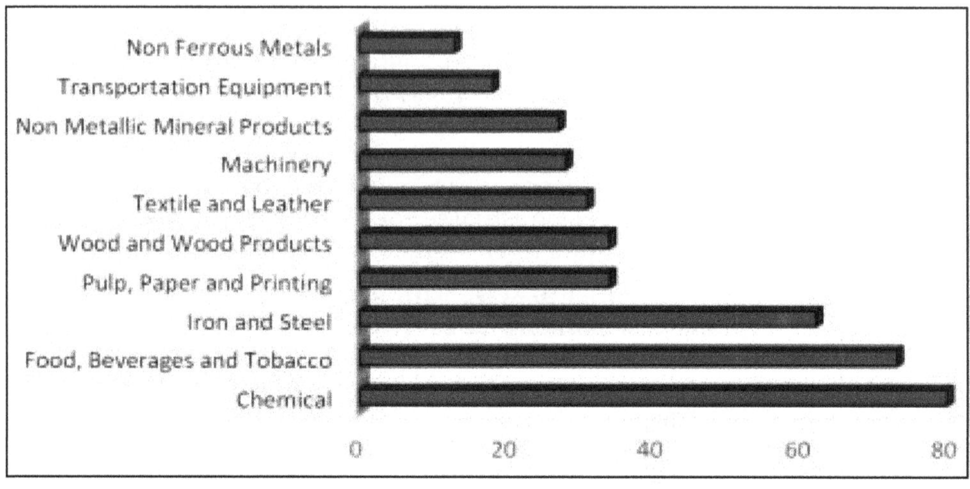

Figure 7.4: Number of Manufacturing Industrial Sub-sectors in Malaysia in 2013.

A simple calculation can be used to generalize the potential impact of solar process heat in Malaysia on fuel's savings and CO_2 emission. Assuming 5 per cent of the industrial heat for temperature below 100 °C were supplied using solar thermal; a 1,577 GWh amount of energy from fuel could be saved which translates into 394,358 tonnes CO_2 emission reduction from fuel oil. This equivalent to 1.8 MWth (405,718 m²) installed capacity. These figures with assumptions used to derive them are shown in Table 7.3.

Table 7.3: Potential Impact of Solar Thermal on Energy Savings from Fuel and CO_2 Emission for Malaysia

Malaysia's Scenario			***Assumption***
Total energy for heating in industrial sector in 2013	105,162,182	MWh	1 ktoe = 11,630 MWh
30 per cent uses heating energy less than 100 °C	31,548,654	MWh	30 per cent of industrial sector uses process heat < 100 °C
Every 5 per cent conversion to solar thermal	1,577,433	MWh	5 per cent heating energy from solar thermal
Reduction of CO_2 emission	394,358	ton CO_2	1 kWh = 0.25 kgCO_2 Emission - Oil
Annual expected energy generation from 4.5kW/m² solar thermal system	3.89	MWh/m²	Industrial practice based on 4 hours daily sun radiation for 270 days a year at 4.5 kW/m² annual and 80 per cent efficiency

Malaysia's Scenario			**Assumption**
Total area collector	405,718	m^2	
Installed capacity of solar thermal	1,826	MWth	1 m^2 = 4.5 kW solar thermal installed capacity

Integration of Solar Heat into Industrial Process

Integration of solar heat into industrial process involves a complex operation than the conventional heat supply system. It requires studying and analyzing the existing heat supply system and determining the potential energy savings, energy flows and temperature levels of the process that could lead to optimizing the economic, technical and energy impact of the system (Cottret, 2010).

There are several ways of solar thermal that can be integrated into the industrial process. First, since the central system for heat supply in industry is providing hot water or steam at a pressure corresponding to the highest temperature needed among the different processes (180 °C - 260 °C), thus, solar systems can be integrated with the conventional heat supply system for preheating water used for processes or for steam generation (Kalogirou, 2003). In this conventional heat supply system, the solar thermal can also be directly coupled to an individual process that has a lower temperature than the central supply (Kalogirou, 2003). These are shown in Figure 7.5.

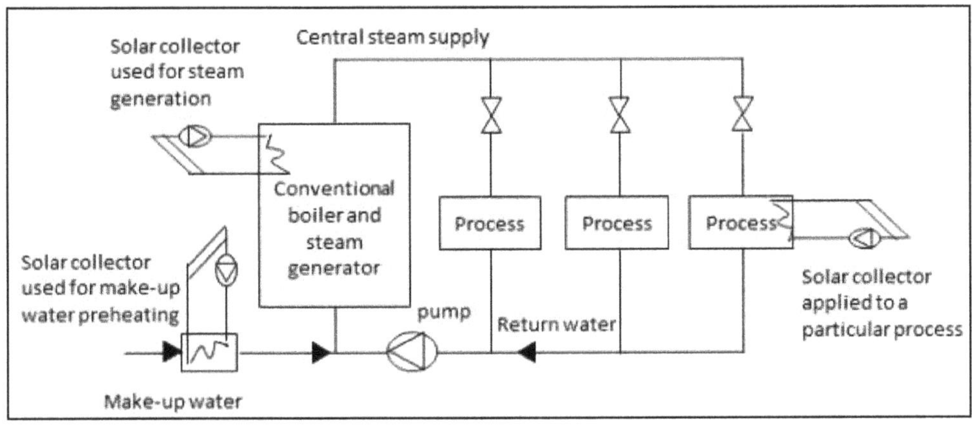

Figure 7.5: Solar Thermal Integration in the Conventional Heat Supply System (Kalogirou, 2003).

Solar thermal can also be integrated directly into existing heating system of the industrial process (Cottret, 2010) as shown in Figure 7.6. For this integration to be working properly, it requires the solar system operates at the same temperature as the existing heating system. Another possible way of integrating solar system in the conventional heat system is to integrate it directly to the process heat (Cottret, 2010) as shown in Figure 7.7. Such integration requires another heat transfer if

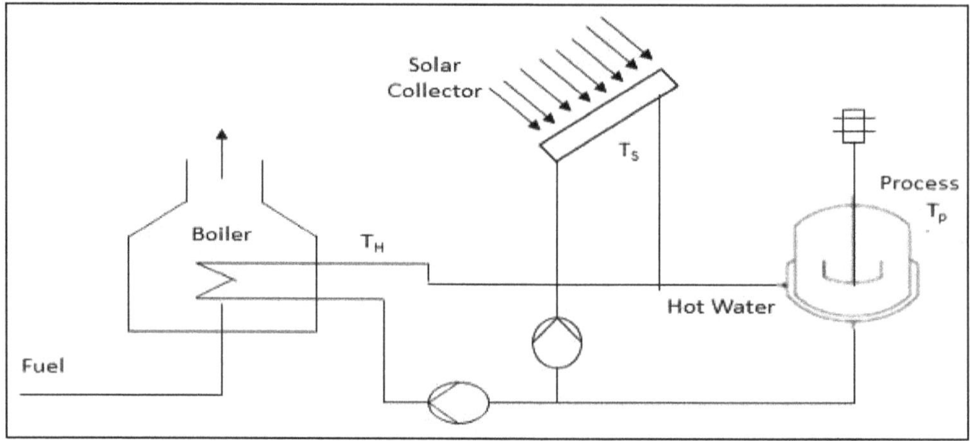

Figure 7.6: Solar Thermal Integration Directly to the Existing Heating System (Cottret, 2010).

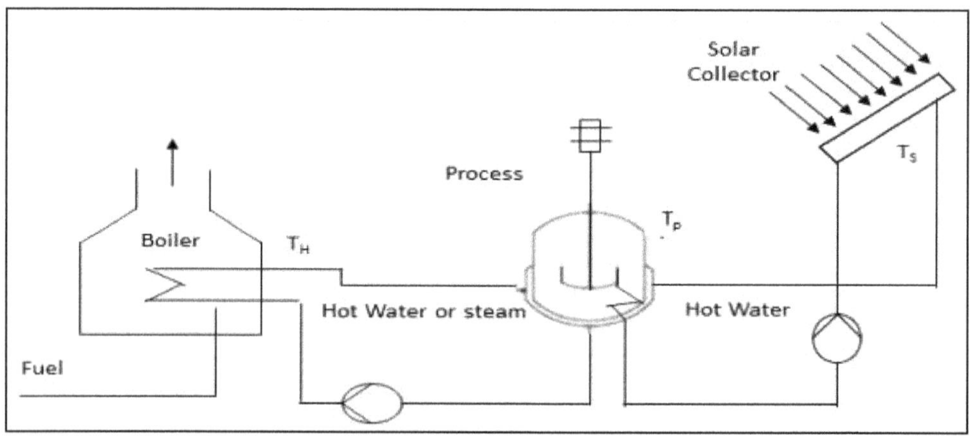

Figure 7.7: Solar Thermal Integration Directly to the Process Heat (Cottret, 2010).

the temperature from the solar collectors is different from the temperature of the heating medium.

A guideline to integrate solar heat into industrial process has been developed by IEA Solar Heating Cooling (SHC) in Task 49 (Muster, 2015). More theoretical analysis on solar heat integration in industry can be found using mathematical model and pinch analyses for fish can industry in (Quijera, 2013; Quijera, 2014) and dairy industry in (Schnitzer, 2007; Quijera, 2011), using static model for textile industry in (Freina, 2013) and using transient analysis for brewery in (Lauterbach, 2014) and dairy industry in (Walmsley, 2015).

4. Demonstration of Solar Thermal Plants for Low and Medium Temperature Process Heat

In this section, four solar process heat plants installed worldwide at various type of industries for low and medium temperature application are presented. These demonstration plants are gathered from the database for applications of solar heat integration in industrial processes created by AEE INTEC (INTEC, 2015).

Fleischwaren Berger Solar Thermal Plant For Meat Industry, Austria

A 746.9 kW_{th} solar thermal capacity has been installed by Fleischwaren Berger GmbH in 2013 for processing and preserving of meat and production of meat products. The installation uses flat plate collector and occupies 106.7 m^2 gross installed collector area in Sieghartskirchen, Austria (Figure 7.8). The produced thermal energy is used for feed water preheating (30 C up to 95 C) for steam production and ham cooking and hot water preheating from about 40 C up to 70 C for drying the air conditioning systems. The point of solar heat integration is at point A1- integration on supply level for heating of make-up water as in Figure 7.9.

Figure 7.8: 746.9 kW_{th} Solar Thermal Plant for Meat Industry in Austria (INTEC, 2015).

Goess Brewery Solar Thermal Plant for Beverage Industry, Austria

A 1,064.0 kW_{th} solar thermal capacity has been installed by Brauerei Goss in 2013 in a beverage manufacturing industry. The solar collector is a flat plate type with gross installed collector area of 1,520 m^2 in Brauhausgasse, Austria (Figure

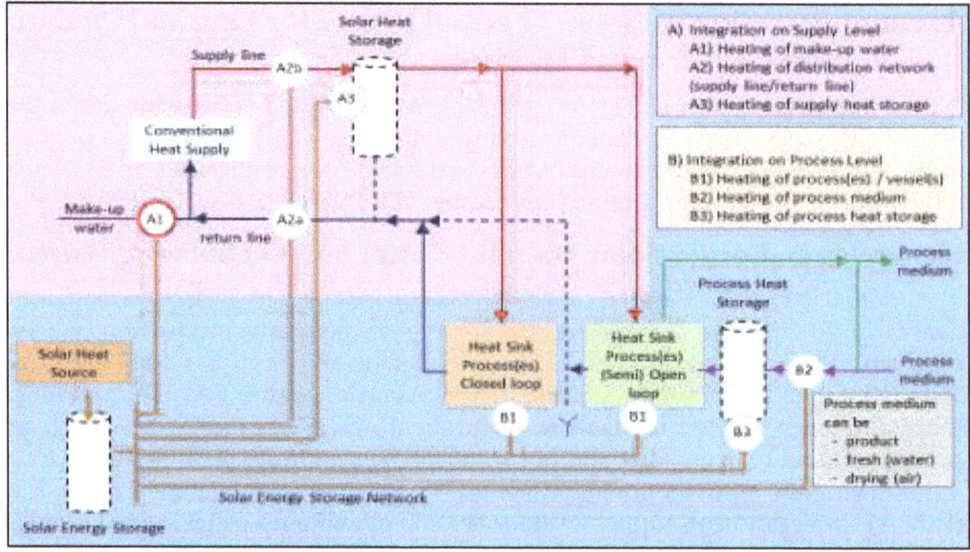

Figure 7.9: Solar Heat Integration Point for Fleischwaren Berger Plant (INTEC, 2015).

7.10). The solar heat integration is in mashing process and make-up water for temperature process between 80 °C to 90 °C. The point of solar heat integration is at point B1- integration on process level for heating of process/vessel as in Figure 7.11.

Figure 7.10: 1,064 kW_{th} Solar Thermal Plant for Beverage Industry in Austria (INTEC, 2015).

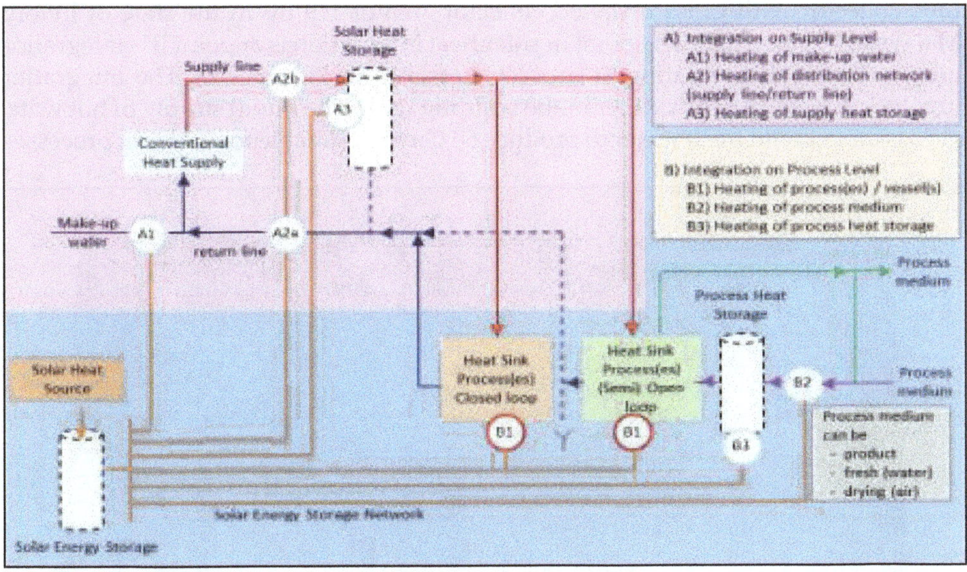

Figure 7.11: Solar Heat Integration Point for Goess Brewery Plant (INTEC, 2015).

PPNJ Poultry and Meat Solar Thermal Plant for Poultry Processing, Malaysia

SIRIM through its SIRIM Fraunhofer program has installed a 83 kW_{th} solar thermal system in 2017 in a poultry processing industry. The system uses evacuated

Figure 7.12: 83 kW_{th} Solar Thermal Plant for PPNJ Poultry and Meat industry in Malaysia.

tube collector with gross installed collector area of 119 m^2 in the state of Johore, Malaysia, (Figure 7.12). The point of solar heat integration is at point B1- integration on process level for heating of process/vessel as in Figure 7.13. The integration involves direct use of hot water in the scalding tank. The initial supply of hot water at 80 is mixed with fresh water to produce 60 Cwater, suitable for scalding processes.

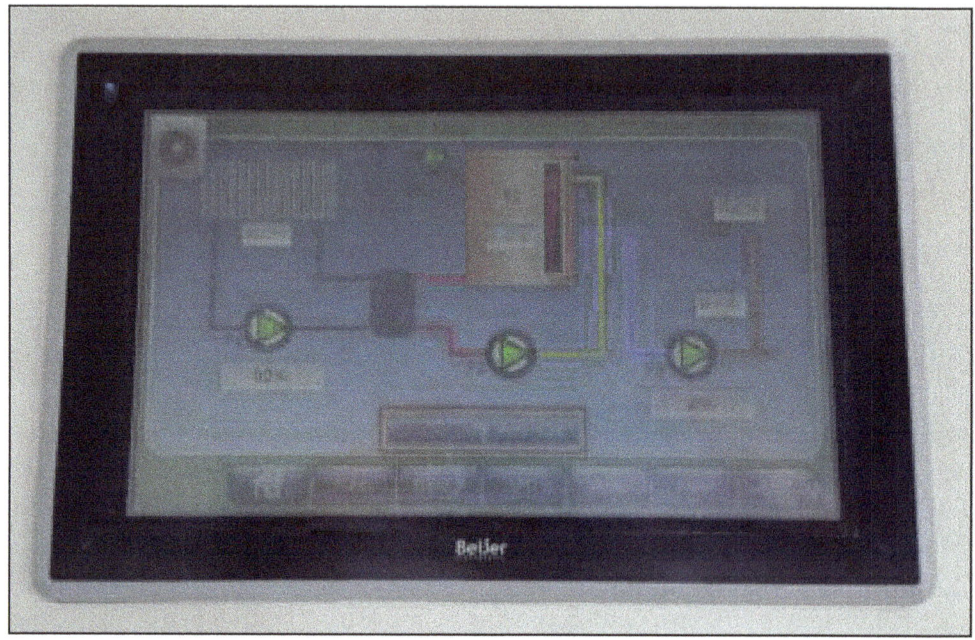

Figure 7.13: Solar Heat Integration Point for PPNJ Poultry and Meat, Malaysia.

De Baron Resort, Langkawi, for Hotel Industry in Malaysia.

SIRIM through it SIRIM Fraunhofer programme has installed a 31.7 kW$_{th}$ in 2017 in one of the hotel chain by De Baron Group of Hotels. The installation uses evacuated tube collector with gross installed collector area of 45.22 m^2 in Langkawi, Malaysia (Figure 7.14). The produced thermal energy is used for providing hot water for bathing to hotel rooms. The integration point is at the process level- shower for hotel room. The temperature of hot water supply is at 60 C and mixed with fresh water to meet individual comfort level.

5. Conclusion

Malaysia has a big potential to implement solar heat for industrial process. This is due to the fact that Malaysia is located at the equatorial region that receives frequent sunlight throughout the year. Furthermore, a significant share of the final energy demand in Malaysia *i.e.* 26.2 per cent in 2013 was used by industries. From this figure, 67 per cent of the energy use in industries is for heating. An enormous impact on the fuel savings and CO_2 emission reduction can be generalised if some

Figure 7.14: 31.7 kW_{th} Solar Thermal Plant for De Baron Resort Hotel, Langkawi, Malaysia.

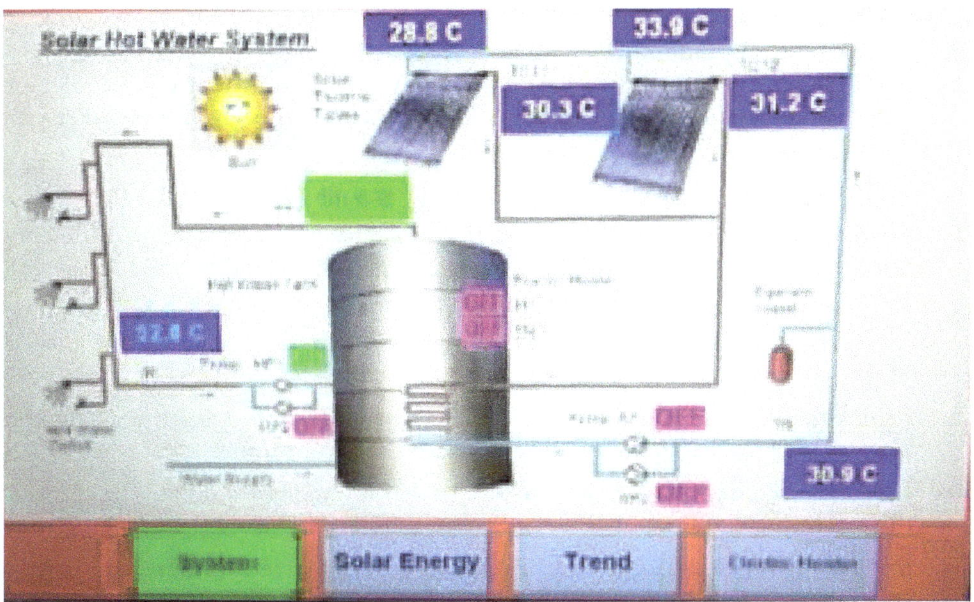

Figure 7.15: Solar Heat Integration Point for De Baron Resort Hotel, Malaysia.

of the shares of the heating demand in industries are supplied by solar energy. From the reviews, low and medium temperature solar process heat are suitable for industrial sectors in Malaysia due to most of its industrial energy demand are from chemical and food industries where most of the processes involve low and medium temperature. Three types of solar collectors can be used for low and medium temperature of process heat namely 1) flate plate collector, 2) evacuated-tube collector and 3) non-tracking and one axis parabolic trough collectors. Solar thermal can be possibility integrated into the industrial process heat using three mechanisms: 1) in the conventional heat supply system (preheating water and steam generation) or directly coupled to individual process, 2) directly to the existing heating system and 3) directly to process heat.

REFERRENCES

1. Affandi, R., Ab Ghani, M. R., Gan, C. K., Jano, Zanariah. (2013). A Review of Concentrating Solar Power (CSP) In Malaysian Environment. *International Journal of Engineering and Advanced Technology (IJEAT)*, 3:378-382.
2. Agency, European Environment. (2015). Final Energy Consumption by Sector and Fuel, in: *016, CSI 027/ENER 016.*
3. Chen, Z., Gu, M., Peng, D. (2010). Heat Transfer Performance Analysis of a Solar Flat-Plate Collector with an Integrated Metal Foam Porous Structure Filled with Paraffin. *Applied Thermal Engineering*, 30: 1967-1973.
4. Cottret, N., Menichetti, E. (2010). Solar Heat For Industrial Processes (SHIP), State of the Art in The Mediterranean Region. Observatoire Méditerranéen de l'Energie.
5. Dagdougui, H., Ouammi, A., Robba, M., Sacile, R. (2011). Thermal Analysis and Performance Optimization of a Solar Water Heater Flat Plate Collector: Application to Te´touan (Morocco). *Renewable and Sustainable EnergyReviews*, 15: 630-638.
6. ECOHEAT. (2006). The European Heat Market, Work Package 1, Final.
7. Ehrmann, N., Reineke-Koch, R. (2012). Selectively Coated High Efficiency Glazing for Solar-Thermal Flat-Plate Collectors. *Thin Solid Films*, 520: 4214-4218.
8. Engel-Cox, J.A., Nair, N.L., Ford, J.L. (2012). Evaluation of Solar and Meteorological Data Relevant to Solar Energy Technology Performance in Malaysia. *Journal of Sustainable Energy and Environment*, 3: 115-124.
9. Faninger, G. (2010). The Potential of Solar Thermal Technologies in a Sustainable Energy Future. IEA Solar Heating and Cooling Programme.
10. Freina, A., Calderonia, M., Mottaa, M. (2013). Solar Thermal Plant Integration into an Industrial Process. *Energy Procedia*, 48: 1152-1163.
11. INTEC, AEE. Database for Applications of Solar Heat Integration in Industrial Processes. [cited 2015; Available from: http://ship-plants.info/?collector_type=4 and from_year=2012.]
12. International Energy Agency (IEA). (2008). Task 33, in: Claudia Vannoni, R.B., Serena Drigo (Ed.), Potential for Solar Heat in Industrial Processes, IEA Solar Heating and Cooling Programme.

13. International Energy Agency (IEA). (2014a). Task 49 Solar Heat Integration in Industrial Processes, 2014 Highlight. IEA Solar Heating and Cooling Programme.

14. International Energy Agency (IEA). (2014b). World Energy Investment Outlook.

15. Kalogirou, S. (2003). The Potential of Solar Industrial Process Heat Applications. *Applied Energy*, 76: 337-361.

16. Lauterbach, C., Schmitt, B., Vajen, K. (2014). System Analysis of a Low-Temperature Solar Process Heat System. *Solar Energy*, 101: 117-130.

17. Lauterbach, C., Schmitt, B., Jordan, U., Vajen, K. (2012). The Potential of Solar Heat for Industrial Processes in Germany. *Renewable and Sustainable Energy Reviews*, 16: 5121-5130.

18. Malaysia, Energy Commission (EC). (2013). National Energy Balance 2013.

19. Mat, S., Ruslan, H. (2015). Solar Thermal Implementation Project for the Industry, in: Malaysia, Launch of the National Project GHG Emissions Reduction in Targeted Industrial Sub-Sector through EE Application of Solar Thermal Systems in Malaysia.

20. Mauthner, F., Weiss, W., Spork-Dur, M. (2015). Solar Heat Worldwide, Markets and Contribution to the Energy Supply 2013. IEA Solar Heating and Coolong Programme.

21. Mekhilefa, S., Saidurb, R., Safari, A. (2011). A Review on Solar Energy Use in Industries. *Renewable and Sustainable Energy Reviews*, 15: 1777-1790.

22. Muster, B., Hassine, I.B., Helmke, A., Heß, S., Krummenacher, P., Schmitt, B., Schnitzer, H. (2015). Task 49 Solar Process Heat for Production and Advanced Applications, Integration Guideline. IEA Solar Heating and Cooling Programme and SolarPACES Annex IV.

23. Quijera, J. A., Alrioz, M.G., Labidi, J. (2014). Integration of a Solar Thermal System in Canned Fish Factory. *Applied Thermal Engineering*, 70: 1062-1072.

24. Quijera, J. A., Garcia, A., Alriols, M. G., Labidi, J. (2013). Heat Integration Options Based on Pinch and Exergy Analyses of a Thermosolar and Heat Pump in a Fish Tinning Industrial Process. *Energy*, 55: 23-27.

25. Quijera, J. A., Alrioz, M.G., Labidi, J. (2011). Integration of a Solar Thermal System in a Dairy Process. *Renewable Energy*, 36: 1843-1853.

26. Schnitzer, H., Brunner, C., Gwehenberger, G. (2007). Minimizing Greenhouse Gas Emissions Through the Application of Solar Thermal Energy in Industrial Processes. *Journal of Cleaner Production*, 15: 1271-1286.

27. Tian, Y., Zhao, C.Y. (2013). A review of solar collectors and thermal energy storage in solar thermal applications. *Applied Energy*, 104: 538-553.

28. Trust, C. (2015). Solar Thermal Technology: A Guide to Equipment Eligible for Enhanced Capital Allowances. Cited 2015; Available from: https://www.gov.uk/government/uploads/system/uploads/attachment_data/file/376187/ECA770_Solar_thermal_technology.pdf.

29. Tutorials, A.E. (2015). Home of Alternative and Renewable Energy Tutorials. Cited 2015; Available from: http://www.alternative-energy-tutorials.com/solar-hot-water/evacuated-tube-collector.html.

30. Vestlund, J., Ronnelid, M., Jan-Olof Dalenback. (2009). Thermal Performance of Gas-filled Flat Plate Solar Collectors. *Solar Energy,* 83: 896-904.

31. Walmsley, T.G., Walmsley, M.R.W., Tarighaleslami, A.H., Atkins, M.J., Neale, J.R. (2015). Integration Options for Solar Thermal with Low Temperature Industrial Heat Recovery Loops. *Energy,* 90: 113-121.

32. Wei, L., Yuan, D., Tang, D., Wua, B. (2013). A Study on a Flat-Plate Type of Solar Heat Collector with an Integrated Heat Pipe. *Solar Energy,* 97: 19-25.

33. Weiss, W., Schweiger, H., Battisti, R. (2015). Market Potential and System Designs for Industrial Solar Heat Applications. AEE INTEC Institute for Sustainable Technologies. Cited 2015; Available from: http://www.aee-intec.at/0uploads/dateien97.pdf.

Chapter 8

Solar Photovoltaic Adoption in Mauritius: Opportunities and Challenges

Vishwamitra OREE

Department of Electrical and Electronic Engineering,
University of Mauritius
E-mail: v.oree@uom.ac.mu

This paper discusses some of the key prospects and challenges for solar PV power development in Mauritius. It elaborates on the barriers to the market deployment of PV solar, without which this renewable energy technology could become competitive with fossil fuels for the generation of electricity. Barriers vary from legal and regulatory to those that are economic, technological and socio-technical in nature. It examines the range of compelling actions that have been taken by the government to remove these barriers so as to foster the uptake of solar PV. It also identifies areas where there is considerable scope for improvement. Given that Mauritius is still at the early stages of solar energy adoption, there exist significant opportunities that the country can exploit to promote solar PV development at a large-scale. They include rapidly falling upfront investments, attracting international financing and technological innovations.

Keywords: *Grid absorption, Barriers, Large-scale adoption, Solar potential.*

1. Introduction

Mauritius, a small island of just 1865 km^2 and with a population of about 1.2 million is located in the South West of the Indian Ocean [1]. Like most small island developing states, the country has no indigenous reserves of fossil fuels and is heavily reliant on imported fossil fuels for its primary energy requirements. In 2015, 84 per cent of the latter was provided by fossil fuels consisting mainly of coal and petroleum products [1]. Besides, primary energy requirements of the country have grown consistently at an annual rate of about 5 per cent during the last ten years, driven by the sustained economic development experienced by the country. During that same period, electricity production has followed a similar trend with the total generation growing by about 27 per cent, from 2320 GWh in 2006 to around 2956 GWh in 2015 [1]. Electricity production in Mauritius relies heavily on steam turbine generators burning fossil fuels as their source of energy. As illustrated in Figure 8.1, diesel and fuel oil along with coal accounted for about 77 per cent of the electricity generated in 2015. Moreover, it can also be observed that bagasse, a by-product of sugarcane, is responsible for a large part of the 23 per cent share of renewable energy (RE) resources in electricity generation. The per capita electricity consumption for the island at 2,148 kWh in 2013 is high compared to other developing nations on the African continent but still low relative to developed countries [2]. The overreliance of the island's electrical power generation on imported fossil fuels is unsustainable for several reasons. Firstly, continuing on the current energy path implies that the stake of energy imports on the national budget is likely to grow as the electricity demand increases in the future. The vulnerability of the country's economy to the volatility of fossil fuel prices was clearly illustrated in 2008 when the hike in petrol prices during the oil crisis caused the import bill to escalate by 28.0 per cent as compared to the previous year [3]. Secondly, the high dependence on fossil fuels, which power most economic and social activities in Mauritius, poses a severe energy security issue. High freighting costs coupled with small market size contribute to the inability of the country to ensure an affordable and continuous supply of energy. These factors further highlight the country's unpreparedness to deal with sudden spikes in fossil fuel prices. Thirdly, the overall greenhouse gas (GHG) emissions of the island are growing at an annual rate of 3 per cent while those resulting directly from the energy sector are increasing at an even higher yearly rate of 5.4 per cent [4]. As a signatory of the United Nations Framework Convention on Climate Change and the Kyoto protocol, Mauritius has committed to work towards reducing GHG emissions and as such, cannot continue along a business-as-usual energy path.

To address this situation, the government pledged to make Mauritius a sustainable island in its long-term energy plan launched in 2009. The blueprint

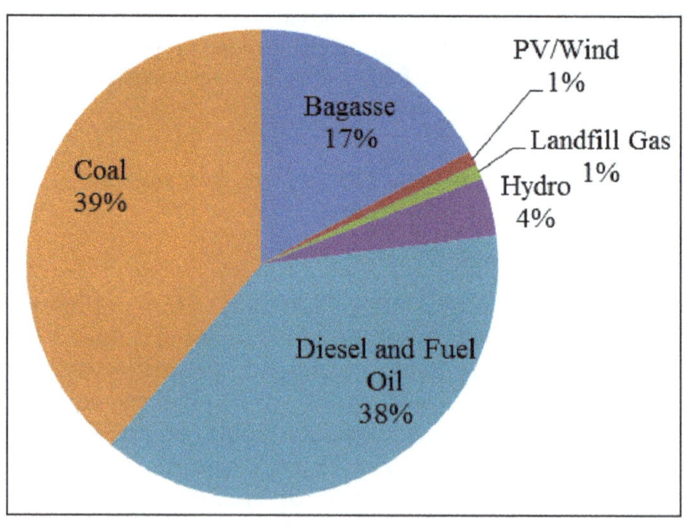

Figure 8.1: Sources of Electricity Generation in 2015 [1].

focused on diversifying the country's energy supply, improving its energy efficiency, addressing environmental and climate changes, and modernising the energy infrastructure [5]. Its highlight was to achieve about 35 per cent self-sufficiency in terms of electricity supply by the year 2025 through the use of RE sources. In this context, a plan was devised to set progressive targets in terms of percentage of total electricity generation from each energy source for successive five-year periods during the planning horizon. According to this scheme detailed in Table 8.1, wind energy is expected to experience the highest forecasted growth until 2025, with a planned share of 8 per cent of the total electricity generated by 2025 compared to a current negligible contribution. In contrast, the target for solar photovoltaics (PV) is deceptively low in view of the enormous potential to harness the abundant solar resource on the island. On the other hand, it is alarming to note that the predicted contribution of coal in the electricity portfolio will drop only marginally so that it is still expected to be the main source of electricity in 2025. Its share is forecasted to decrease from 43 per cent presently to only 40 per cent in 2025. Thus, this energy plan will continue to lock in Mauritius on coal for its electrical power requirements in the foreseeable future along with all its associated environmental, economic and social implications. It is felt that given the promising renewable potential of the island, any energy plan should focus on enhancing the exploitation of readily available RE sources, predominantly solar.

Table 8.1: Forecasted Electricity Generation by Energy Source for the Period 2010-2025 [4]

Fuel Sources		*Percentage of Total Electricity Generation (per cent)*			
		2010	*2015*	*2020*	*2025*
Renewable	Bagasse	16	13	14	17
	Hydro	4	3	3	2
	Waste to energy	0	5	4	4
	Wind	0	2	6	8
	Photovoltaic	0	1	1	2
	Geothermal	0	0	0	2
	Sub-total	20	24	28	35
Non-Renewable	Fuel Oil	37	31	28	25
	Coal	43	45	44	40
	Sub-total	80	76	72	65
	Total	100	100	100	100

The rest of the paper discusses some of the key prospects as well as the challenges for large-scale solar PV power in Mauritius and is structured as follows: Section 2 examines the solar potential in Mauritius. Section 3 provides an overview of the progress that has been made on the solar PV front in terms of policy decisions and projects implemented. Section 4 discusses the barriers that are hindering the exploitation of the promising solar potential of the country. Section 5 elaborates on the challenges that Mauritius faces to promote solar PV further as the legislation on carbon emissions become increasingly stringent and uncertainty over the supply and price of fossil fuels escalate. Finally, Section 6 looks at the opportunities that can contribute to the setting up of a conductive environment for large-scale solar PV integration.

2. Solar Potential in Mauritius

Tropical islands are known for their long hot summers together with abundant sunshine and Mauritius is no exception. Long-term data recorded by the Mauritius Meteorological services reveal that the island receives between 2,350 to 2,850 hours of bright sunshine annually depending on the location, as depicted in Figure 8.2 [5]. Mauritius therefore gets much more sunshine than Germany, the world's leading country for per capita solar PV deployment, which receives between 1450 and 1700 hours of bright sunshine annually [6]. Besides its environmental benefits, solar PV also has numerous other advantages. Unlike wind turbines, it does not involve noise and poses no risk to birds. The long lifetime of solar PV panels, typically between 20 and 30 years, promises good returns on the initial investment in the long term. Since solar PV systems have moving parts and negligible wear, they require minimal maintenance. Furthermore, solar PV systems are modular so that their capacity can vary from a few Watts to MW. They can also be implemented as a distributed system independent of the grid or as a grid-connected system. It is surprising then, that in spite of such huge potential for solar power and its many benefits, the share of solar PV in the energy mix is insignificant.

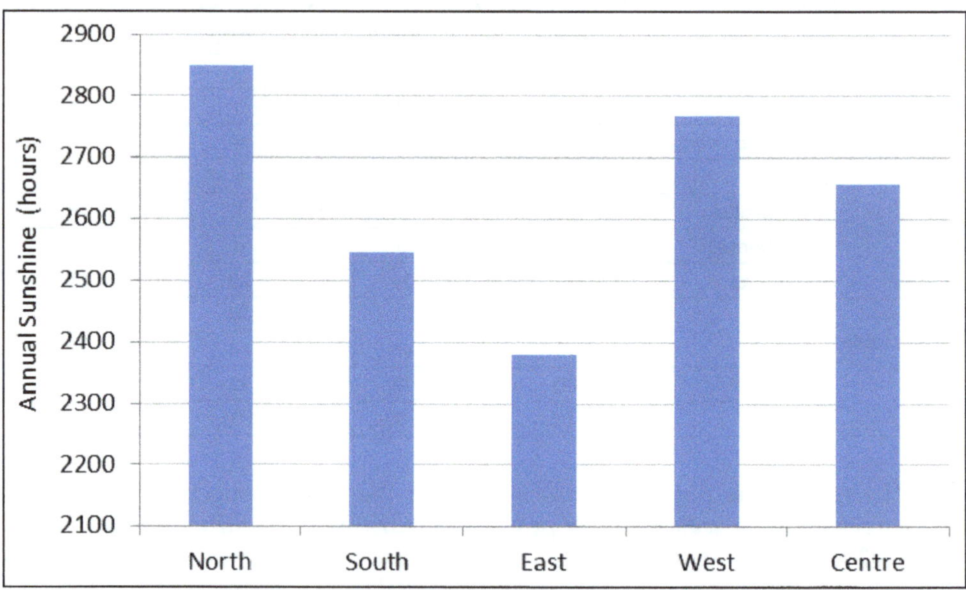

Figure 8.2: Average Annual Sunshine Hours at Various Regions in Mauritius [5].

3. Barriers to Solar Energy Deployment

The main challenge faced by many small islands like Mauritius in order to fully exploit their abundant solar resources is to find ways to convert this RE potential into an energy form that is accessible, affordable and adaptable. However, considerable barriers to the wide-scale deployment of solar energy technologies were identified during the early stages of the long-term energy transition of Mauritius. They are described hereunder [7].

- ✰ The major problem for Mauritius was the absence of appropriate regulatory, technical and market conditions required to foster the implementation and operation of solar energy projects, particularly, grid-connected PV projects. To start with, the governmental authorities responsible for formulating and implementing the government's energy policy did not have appropriate legal provisions that would facilitate access to new entrants in the exploitation of RE sources for electricity generation. The Ministry of Energy and Public Utilities (MEPU) is responsible for formulating policies and overseeing the legal framework in the energy sector while the Central Electricity Board (CEB) is the national electricity utility which operates under the purview of the MEPU. Prior to 2010, regulations dealing with RE existed as a disparate set of laws. There was also no appropriate licensing structure for solar PV installations. In this context, the introduction of a comprehensive market-oriented energy policy together with its accompanying legal and regulatory framework was crucial to expand solar power beyond a niche market. This would help in creating appropriate market signals to induce investment in solar energy.
- ✰ High upfront costs represented another key stumbling block for the development of solar energy projects. Although many solar technologies in general and PV in particular, are well-established with rapidly decreasing costs, comparatively small project sizes on islands can often restrain funding opportunities. It is therefore necessary to come up with various incentives for investors as well as ownership models and feed-in tariffs to overcome the financial barrier.
- ✰ Several technical and infrastructural issues were also identified as major inhibitors to the widespread adoption of solar energy. A PV system consists of the PV module along with the balance of system components which usually include a transformer, an inverter, switches, wiring and racks. Guidelines and technical standards are required to ensure the smooth connection of such systems to the grid. Moreover, like most electricity systems worldwide, the Mauritian power system was initially designed to accommodate conventional generation technologies. The intermittent nature of the solar resource makes it non-dispatchable unlike fossil fuel-based generation. Thus, integration of RE resources in the existing grid beyond a certain level calls for major upgrading of the existing grid infrastructure. The absence of a high quality radiometric network that would enable an accurate mapping of the solar resources in Mauritius

constitutes yet another major technical barrier. Existing pyranometers of the Mauritius Meteorological Services are not in working condition. The spatial distribution of solar resources at a small-scale over the country represents a valuable tool for guiding potential PV investors in making informed decisions.

- Finally, the lack of awareness among the public was recognised as a hurdle in the uptake of solar energy in Mauritius. Any attempt towards integrating substantial amount of RE in the grid is likely to fail if the support of the population at large is not garnered in the process. Yet, when the energy transition was conceived in 2009, a large segment of the population was not fully aware of the environmental and financial benefits that solar PV systems could bring to them. The common perception among the public was that solar PV systems are costly alternatives for electricity generation. Several factors contributed to this problem including novelty of the solar PV concept in the island, lack of human capital locally to plan, install and maintain such systems, shortage of reliable data from existing solar energy projects that would confirm their economic viability, and best practices derived from past projects.

4. Solar PV Projects

A major project entitled "Removal of Barriers to Solar PV Power Generation in Mauritius, Rodrigues and the Outer Islands" was funded by the United Nations Development Programme (UNDP) under the Global Environment Facility (GEF). It aimed at assisting the government in addressing the barriers to solar PV markets in Mauritius [8]. Once the main barriers to the large-scale adoption of solar power in Mauritius were identified, government initiated tangible actions to mitigate them. Several policy initiatives were taken to stimulate the electricity sector to shift from conventional energies to environmentally sustainable technologies and to guarantee fair access to the electricity market. On the technical front, a Grid Code was established in December 2010 to define the technical criteria and requirements for connecting small-scale solar PV systems to the low-voltage network of the national grid [9]. In parallel, a Small-Scale Distributed Generation (SSDG) scheme was introduced to enable small-scale electricity producers to generate electricity for their usage and sell any excess generated power to the grid at preferential rates guaranteed for 15 years. These rates ranged from MUR 25, 20 and 15 per kWh for micro (up to 2.5 kW), mini (up to 10 kW) and small (up to 50 kW) installations respectively [9]. A demand-side management measure was also included in the feed-in tariff to urge SSDG owners to synchronise their electricity consumption with the generation. Thus, if the system owner did not consume at least a third of the total electricity generated by the system, the rates for the following year would be reduced by 15 per cent. It was imperative to offer attractive rates, well above the normal rates of MUR 3 to 8 per kWh paid by consumers, to kick-start the solar PV market development given the prohibitive upfront investment. In the absence of these highly subsidised rates, the payback period of a small-scale solar PV system would exceed its expected lifetime.

Initially targeting a cumulative capacity of 2 MW, the appeal of the scheme to the public was such that a second phase of the SSDG scheme was required in December 2011 to bring the total capacity to 3 MW [9]. Alongside, an additional 2 MW SSDG scheme was initiated for public, educational, charitable and religious institutions to boost interest in RE among non-profit making organisations. The tariff for the surplus electricity sold to the grid was based on the average marginal cost of electricity production of the CEB as the primary goal was to provide an opportunity for such institutions to decrease their electricity bill through sponsored RE systems [10]. A Grid Impact study performed by the CEB after the first 200 SSDG installations were commissioned revealed that the addition of these intermittent inputs did not have a significant impact on the stability and reliability of the electrical network [11]. By the end of 2014, 293 solar PV SSDG systems representing a total capacity of 2,622 kW were already in operation [12].

In the wake of the vastly successful SSDG scheme, a Medium-Scale Distributed Generation (MSDG) scheme was launched to cater for the interest of several promoters to set up RE systems beyond 50 kW capacity. To this end, two Grid Codes were established to define the technical requirements for connecting RE systems of capacities in the range of 2 to 50 MW and 50 to 200 MW to the 22 kV medium-voltage network. MSDG tariffs are determined on a case-to-case basis and take into account the cost of grid integration as well as the cost of supply of electricity to the consumer during no sunshine periods if the system is not equipped with battery storage [13]. Following a request for proposal, five Energy Supply and Purchase Agreements (ESPA) were signed with preferred bidders in 2014, each for solar PV systems of 2 MW [12]. In August 2015, the third phase of the SSDG scheme was launched for an anticipated capacity of 5 MW. Having achieved its main objective of motivating the public to embark on the energy transition, the feed-in tariff was dropped in favour of net-metering. The latter allows SSDG owners to offset the cost of electricity drawn from the grid with credits earned when their system exports surplus electricity to the grid.

On a larger scale, the first solar PV farm started its operations in February 2014. This 15.2 MW solar PV plant was implemented at the foot of a mountain in Bambous, near the Eastern coast of the island. It consists of 60,800 PV panels covering an area of 34 hectares. Until December 2014, it had generated about 20 GWh of electricity, accounting for about 0.8 per cent of the total units generated during that year [12]. A new 66 kV bay was also constructed to allow the connection of the farm to the high-voltage electrical network.

5. Challenges

Although some noteworthy strides have been made to motivate RE adoption in Mauritius, large-scale expansion of solar PV continues to face significant challenges. Appropriate regulation has the potential to tackle these challenges. The legal and institutional framework governing the energy sector in Mauritius is still fraught with regulatory deficiencies [14]. The green revolution is radically changing the traditional landscape of the electricity sector. New paradigms are emerging in the generation, transmission and distribution processes, spurring the adoption of new policies and regulations. The latter must encourage investment in the power sector and focus on energy efficiency, design of rational tariffs and laws to foster RE technologies. It is also essential that an independent regulator monitors and oversees the energy market to balance the interests of different stakeholders. In Mauritius, the CEB wears several hats as it is not only the sole owner and operator of the transmission and distribution systems of the national electricity network, but also the leading power supplier and sole regulator of the power sector. It is high time to unbundle this highly integrated utility model into separate entities in order to eliminate inefficient monopolies and promote fair competition on the electricity market.

With the growing number of solar PV and wind turbine installations, there was a pressing need to assess the impact of the increased integration of intermittent RE on the stability and reliability of the electrical network. The unpredictable and variable nature of solar and wind energy imply that the magnitude and frequency of the fluctuations in the generated power increase as their share in the energy mix rises. The particular conditions of small islands exacerbate this problem. For example, geographical isolation means that any excess or shortage of generation from the intermittent sources cannot be exchanged with neighbouring power grids. The rapid fluctuations in intermittent RE output as meteorological conditions change require the operation of expensive alternative power sources that have the ability to respond quickly. The electricity network in Mauritius presently can only deal with a relatively small degree of variable RE integration. High investment in additional infrastructure is needed if the level of integration is to be increased further.

Another key challenge was identified during the implementation of the SSDG scheme when it was found that there is a dearth of trained staff having the appropriate skills and know-how to properly size, install, operate and maintain solar PV systems. As a result, the existing small pool of technical staff was overwhelmed by the demand during the implementation phase of the scheme [14]. Finally, the high capital costs of solar PV systems imply that they are still unaffordable for the majority of the population.

6. Opportunities

Amidst all these challenges, there also exist new windows of opportunities to sustain the deployment of solar PV at a scale needed to alleviate the overdependence of Mauritius on fossil fuels. Solar PV global capacity has maintained an exponential growth during the last two decades such that it has now positioned itself as one of the most promising technologies to support the transition to a decarbonised energy supply [15]. The remarkable decline in the prices of solar PV cell cost due to economies of scale and their manufacture in China has been one of the main drivers of this shift. The average solar PV system costs have plummeted by 82 per cent in the last six years alone [16]. At the same time, system efficiencies have improved considerably therefore providing more power for the same capacity. In light of the plummeting costs, enhanced outputs and long warranties, it is expected that in the future, solar PV systems may be able to successfully compete with fossil fuels in sunny regions with high electricity prices without subsidies [17].

To anticipate the technical issues regarding integration of intermittent RE in the grid, the CEB initiated a grid absorption capacity study in 2014 with the support of UNDP and the World Bank. The study showed that grid stability was already a critical concern given the total number of intermittent RE projects in the pipeline [14]. Nevertheless, the study proposed cost-effective and centralised solutions to mitigate the problem. The main recommendations included enhancing the existing generation control systems as well as the installation of an Automatic Generation Control (AGC) system to improve the power system frequency control, increase the storage capacity within the electrical network through the installation of lithium-ion batteries combined with grid-edge solutions, and upgrading the power system control with accurate RE generation forecasting tools [14]. It is expected that implementation of these measures at a total cost of around US$ 35.3 million will result in an increase of the intermittent RE integration capacity of the grid from its current limit of 60 MW to about 185 MW [14]. It is likely that this new threshold of integration capacity will catalyse investment in PV solar.

With this in mind, the Green Climate Fund (GCF) has recently approved the funding of a proposal entitled "Accelerating the Transformational Shift to a Low-Carbon Economy in the Republic of Mauritius" submitted by the UNDP to the tune of US$ 28.2 million. The total estimated value of the project is US$ 191.4 million, with the major stake of US$ 123.9 million coming from the Mauritian government and the rest from a loan by the Agence Francaise de Development (AFD) and a grant from UNDP [18]. Besides expanding the grid absorption capacity, the other major aspect of this project concerns the institutional strengthening for RE. In this regards, the project intends to provide the MEPU with the necessary assistance to develop the appropriate legal and regulatory framework to allow uptake of RE [14]. In addition, it proposes to reinforce and empower the Mauritius Renewable Energy Agency (MARENA), which was set up in 2016 to oversee the development of RE in the country. Ultimately, MARENA will be able to assist potential RE investors by decreasing transaction costs and time delays. MARENA is also expected to provide technical oversight and policy planning support [14]. Lastly, the project will implement capacity building by ensuring technical training on the installation, operation and maintenance of solar PV systems.

7. Conclusions

The challenge of diversifying the energy mix of the electricity sector in Mauritius by integrating more intermittent RE sources is huge. Steps have been taken to put Mauritius on course to achieve its energy transition. Nonetheless, much remains to be done. Considerable changes have to be brought in the policy, regulatory, and institutional frameworks in order to create a conducive environment to nurture the uptake of solar PV. Meanwhile, substantial investments are required to upgrade the grid so that it can absorb large quantities of variable power. Accompanying measures like public awareness campaigns and training of sufficient human capital to acquire the necessary skills and know-how to effectively plan, install and maintain solar PV systems. Mauritius has been successful in mobilising funds and knowledge transfer from international institutions devoted to the promotion of RE technologies such as GCF, UNDP, AFD and GEF.

REFERENCES

1. Statistic Mauritius, 2016. Energy and Water Statistics - Year 2015. Republic of Mauritius. Available at http://statsmauritius.govmu.org/English/Publications/Pages/Energy-and-Water-Statistics-Year-2015.aspx
2. The World Bank, 2016. Electric power consumption (kWh per capita). Available at: http://data.worldbank.org/indicator/EG.USE.ELEC.KH.PC
3. Ministry of Environment and National Development Unit, 2010. Mauritius Strategy for Implementation National Assessment Report 2010.
4. UNFCCC, 2010. Second National Communication of the Republic of Mauritius under the United Nations Framework Convention on Climate Change. Available at: http://unfccc.int/resource/docs/natc/musnc2.pdf. Accessed on 8 March 2017.
5. Mauritius Meteorological Station, 2017. Climate of Mauritius. Available at: http://metservice.intnet.mu/climate-services/climate-of-mauritius.php. Accessed 04 March 2017.
6. Ministry of Renewable Energy and Public Utilities, 2009. Republic of Mauritius Long term Energy Strategy 2009-2025.
7. Current Results weather and science facts: Average Sunshine a Year in Germany. Available at: https://www.currentresults.com/Weather/Germany/annual-hours-of-sunshine.php, Accessed on 5 March 2017.
8. Global Environment Facility, 2011. Mauritius: Removal of Barriers to Solar PV Power Generation in Mauritius, Rodrigues and the Outer Islands. Request for CEO endorsement/approval.
9. Central Electricity Board, 2010. Grid Code for Small Scale Distributed Generation. Available at: http://ceb.intnet.mu/grid_code/document/gridc.pdf. Accessed on 7 March 2017.
10. Central Electricity Board, 2013. Integrated Electricity Plan 2013-2022. Available at: http://ceb.intnet.mu/CorporateInfo/IEP2013/Chapter5_Power/Generation/Plan.pdf. Accessed on 7 March 2017.
11. Busgeeth, R., 2014. Small-Scale Distributed Generation Project. Available at: http://publicutilities.govmu.org/English//DOCUMENTS/SSDG/PRESENTATION.PDF. Accessed on 8 March 2017.
12. Central Electricity Board, 2014. Annual Report 2014: Power supply challenges. Available at: http://ceb.intnet.mu/CorporateInfo/ar2014.pdf. Accessed on 8 March 2017.
13. Hansard No. 8 of 2016. Sixth National Assembly Parliamentary debates. Available at: http://mauritiusassembly.govmu.org/English/hansard/Documents/2016/hansard0816.pdf. Accessed on 8 March 2017.
14. Green Climate Fund, 2016. Consideration of funding proposals – Addendum VI Funding proposal package for FP033. Accelerating the Transformational Shift to a Low-Carbon Economy in the Republic of Mauritius.

15. Duan, H. B., Zhu, L. and Fan Y. (2014). A cross-country study on the relationship between diffusion of wind and photovoltaic solar technology. Technological Forecasting and Social Change, Vol. 83, pp. 156–169.
16. EY, 2016. Capturing the sun: The economics of solar investment.
17. Kurtz, S., Atwater, H., Rockett, A., Buonassisi, T., Honsberg, C., Benner, J., 2016. Solar research not finished. Nature Photonics, Vol. 10(3), pp.141-142.
18. Green Climate Fund, 2017. Project FP033 - Accelerating the Transformational Shift to a Low-Carbon Economy in the Republic of Mauritius. Available at: http://www.greenclimate.fund/-/accelerating-the-transformational-shift-to-a-low-carbon-economy-in-the-republic-of-mauritius. Accessed on 9 March 2017.

Chapter 9

On Morocco's Renewable Energy After COP 22

Ismail Mekkaoui Alaoui

Physics Department, Faculty of Sciences Semlalia,
Cadi Ayyad University, BP 2390 Marrakech, Morocco
E-mail: mekkaoui@uca.ac.ma

Solar photovoltaic and thermal conversion systems are suitable for most of Morocco's territory. In some areas, solar energy can reach 7.5 kWh/m^2/day during May-September period. Morocco, the host country of COP22, is developing a policy roadmap to transition to 100 per cent electricity from renewable energies. In 2015, (during the COP21) Morocco announced its goal to raise the share of renewable electricity to 52 per cent by 2030 (20 per cent solar, 20 per cent wind and 12 per cent hydro). To reach this goal, the country will develop additional electricity production capacity between 2017 and 2030 of around 10,000 MW in renewable energies: 4,560 MW solar, 4,200 MW wind and 1,330 MW hydro. Concentrator Photovoltaics (CPV) is one of the most promising technologies to produce solar electricity at competitive prices. High performing CPV systems with efficiencies well over 30 per cent and multi-megawatt CPV plants are now a reality. For instance, Noor Ouarzazate (east central Morocco), the first solar mega-project is a good example. It will reach a total capacity of 580 MW by 2018.The Tarfaya's (South east of Morocco) wind park, with a production potential of 1,084 GWh/year, which is already supplying 1.5 million households and has become Africa's largest wind energy project.Indeed, in Morocco renewable energy is not only a very important factor for the environment and the production of goods and services, but a key development vector. In this paper, we will present the Morocco's renewable energy potential and main installations, with emphasis on the CPV (Noor1-Noor3). An overview of this technology including: the electrical characterization of photovoltaic solar cells and some aspects of the transition to 100 per cent electricity from renewable energies will be discussed.

Keywords: *Renewable energies, Electricity, Concentrator photovoltaic, Solar, Wind, Electrical characterization.*

1. Introduction

Conversion of solar energy into electricity is now and will be widely used in Moroccan dry and semi-dry areas. In the south and east of Morocco (Tafilalet-Draa region) the sun is usually shining for the most of the year. In this region are (will be) implemented most of the photovoltaic installations. The number of the wet days in these areas does not exceed two or three days per month during autumn, winter and spring; while in summer the cloudy days are rare. It may happen, during dry conditions that these regions do not see a cloudy day for few months. The temperature may reach 50 °C in some area in July and August during day time. In these conditions photovoltaic systems are very efficient for many applications [1]. For instance water is very scarce in these regions but the underground water can supply some of the needs using solar photovoltaic pumping systems [2-3]. Table 9.1 shows the average insulation, the clearness and the precipitations in the Tafilalt-Draa region. These data were obtained from NASA Langley Research Center Atmospheric Science [4]. The data showed in this table can be extended to the most of the southeast of Morocco.

Table 9.1: Average Insulation and Precipitation in Tafilalt-Draa Region

Month	*Insulation (KWh/m²/day)*	*Clearness*	*Precipitation (mm)*
Jan	2.84	0.52	16
Feb	3.92	0.57	9
Mar	4.94	0.58	14
April	6.17	0.61	20
May	6.75	0.61	13
June	7.28	0.64	3
July	7.22	0.64	1
Aug	6.4	0.61	6
Sep	5.37	0.59	11
Oct	4.20	0.57	21
Nov	3.09	0.53	17
Dec	2.48	0.49	12

Morocco, being one of the largest energy importer in MENA, is making concerted efforts to reduce its reliance on imported fossil fuels [5]. Renewable energy is an attractive proposition. Annual electricity consumption in Morocco was 33.5 TWh in 2014, and is steadily increasing at a rate of around 3 per cent each year (in 2015, 34,4TWh) [6-7]. The major sources of alternative energy in Morocco are solar and wind. Wind energy potential is excellent in vast parts in the northern and southern regions, with the annual average wind speed exceeding 9 m/s at 40 meters elevation. Wind systems are suitable in many places due to the regularity and relatively high wind speed. As far as solar is concerned, the country experiences 3000 hours per year of annual sunshine equivalent to 5.3 kWh/m /day.

Morocco has launched in 2011 an ambitious program to provide 20 per cent of its electricity from PV in 2020. In this development, advantages are beginning to be shared by society as a whole. For example in NOOR Ouarzazate, the first project launched by the Moroccan Solar Energy Agency (MASEN), covering 3,000 hectares with a total capacity of 580MW by 2018 [8]. Huge Potential is available, CPV are located in desert/high intensity/long sunlight hours. Parabolic mirrors reflect/focus sun's rays onto metal water pipe located along focal axis of mirrors. High temperature produced steam which is used to electrical power generated [9].

The dam's politics launched in early sixties of last century helped in providing electricity from hydro potential, even though a major part of the Moroccan climate is arid or semiarid. The total installed renewable energy capacity (excluding hydropower) was approximately 787MW at the end of year 2015. The Moroccan Government has set up an ambitious target of meeting 42 per cent of its energy requirements using renewable resources (2GW solar and 2GW wind) by 2020 [10-11]. During the COP21 Morocco announced its goal to raise the share of renewable electricity to 52 per cent by 2030 (20 per cent solar, 20 per cent wind and 12 per cent hydro). To reach this goal, the country will develop additional electricity production capacity between 2017 and 2030 of around 10,000 MW in renewable energies: 4,560 MW solar, 4,200 MW wind and 1,330 MW hydro [12-13].

2. Charaterization of Solar Cells and Panels

The models of the current voltage characteristic of an illuminated solar cell generator can be described by implicit equations related to a Single Exponential Model (SEM) or to a Double Exponential Model (DEM) [14]. Different sets of computer parameters can be found from numeric procedures for experimental characteristic measured under specified illumination and temperature conditions [15]. A criterion of effectiveness is defined to evaluate the performance of a set of computer parameters and helps to derive the best-descriptive set of parameters. These are related to the physical parameters and their variations are established from dark to light concentration. The double exponential model can produce a good description of the experimental characteristic. The percentage deviation between the calculated and measured currents is less than 1 per cent for the used samples.

2.1 Characterization Model

Under illumination and normal operating conditions the photovoltaic solar cell behaves like a generator producing a current I and a voltage V. For a specified illumination I_{ph} and temperature T, its characteristic can be described by the implicit equation [16]:

$$I = I_{ph} - \frac{V + IR_s}{R_{sh}} - I_1\left(\exp w - 1\right) - I_2\left(\exp(\frac{w}{n}) - 1\right) \quad (equ.1)$$

where, $w = q\frac{V + IR_s}{KT}$

The six parameters to be evaluated numerically are: R_s (series resistance) ; R_{sh} (shunt resistance); n (quality factor); I_1 (ideal current of diffusion); I_2 (current due to the recombination via deep levels in the space-charge region of the junction); and I_{ph} (photo-current). This equation suggests an equivalent circuit, Figure 9.1, the double exponential model (DEM),

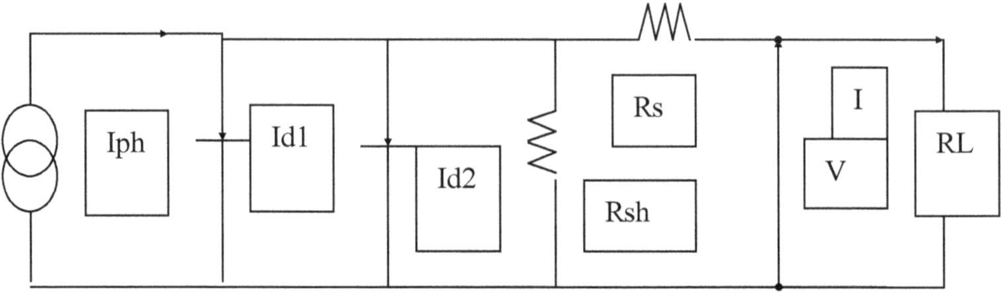

Figure 9.1: Double Exponential Models for *I-V* Characterization of Solar Cells.

2.2 DATA Acquisition System and Samples

The data-acquisition system is described elsewhere [14-16]. Numerical methods that make use of the experimental data for three points: I_{sc}, V_{oc}, and the maximum power point (I_{mp}, V_{mp}), and around these points are explored. The result of this procedure is an exact correspondence between the calculated and experimental characteristics for and around these points and consequently yield the same fill factor and efficiency. For each characteristic, under specified conditions of illumination and temperature, the parameters I_{ph}, I_{01}, I_{02}, n, R_{sh} and R_s, are calculated. We consider these sets of parameters as computer parameters to determine the set of parameters which can provide the best possible description of the characteristic: these particular values will be considered as "descriptive parameters" which are expected to be related to the physics underlying the electrical characteristics of the photovoltaic solar cell. They can be studied for illuminations ranging from dark (I_{ph}= 0) to light concentration. The method has been applied successfully to solar cells made from polycrystalline Si and might be used for any solar cell which can be characterized by a double exponential models.

2.3 Criterion of Effectiveness of a Set of Computer Parameters

Having determined the values of a set of computer parameters, we want to evaluate its effectiveness. The current intensity value I and the dynamic resistance R are calculated for each experimental value *V*(*N*) from equation (1). Each measurement [*V*(*N*), *I*(*N*)] is then associated to a calculated point *[V*(*N*),*I]*. The deviations *[I*(*N*)-*I]*/*I*(*N*) would be minimized in the short-circuit area and maximized in the open-circuit area, and reversely for deviations calculated on the voltage. We are then led to consider the distances d(N) between the calculated characteristic and the experimental points [16] to allow comparisons.

$$d(N) = \frac{I - I(N)}{I_{SC}} \sin(tg^{-1} a) \qquad (equ.2)$$

With,

$$\tan a = -\frac{dV}{dI}\frac{I_{SC}}{V_{OC}} = R\frac{I_{SC}}{V_{OC}} \qquad (equ.3)$$

The root-mean-square of these distances is then calculated for the experimental points such as $V(N) > 0$:

$$Q = \left(\frac{1}{L}\sum_{N=1}^{L} d(N)^2\right)^{1/2} \qquad (equ.4)$$

A perfect description of the experimental points by a set of computer parameters would yield $Q=0$; reversely, the further the calculated curve is from the points, the higher Q becomes. The effectiveness of a model is optimum where Q is minimum.

We need to mention also that photovoltaic cells and consequently photovoltaic panel suffer from a drop in efficiency with the rise in temperature due to increased resistance [17]. But the rise in temperature can increase the efficiency by contributing to the liquid heating in the thermal solar conversion in the projects Noor Ouarzazate. CPV modules use multi-junction cells. These cells are actually comprised of several solar cells, called sub-cells, connected in series and made of different semi-conductor materials. The advantage of this configuration is that it allows for greater use of energy from the photons present in the solar spectrum, thus providing the system with a higher degree of conversion of light into DC electricity. The adequacy of the DEM to describe the I-V characteristic is good mainly for single solar cells, but for modules we need to take into account other factors [18], such as the manufacturer data sheet, the inclination, *etc.*

3. Morocco's Renewable Energies Potential

3.1 Morocco's Renewable Energy main Locations

Morocco is located between two climates (Mediterranean in the north and deserted in the south) and hence benefits from a topography and climate which are extremely well suited to the development of renewable energies, in particular solar and wind. The hydraulic part depends mainly on the rain and the dams, the main dams producing electricity are concentrated in the North West. The solar radiation is intensive and tremendous in all part of Morocco and mainly in the east and south. The country is having one of the highest rates of solar insolation (3,000 h/year of sunshine, up to 3,600 h/year in the desert). Wind potential is located in some Morocco's coastal regions, specifically in top north west and the southern region where there is large and steady wind system with the annual average wind speed exceeding 9 m/s at 40 m elevation.

3.2 Solar and Wind Renewable Energy main Sites

The major sources of alternative energy in Morocco are solar and wind, Wind installations for electricity systems are/will be located in the northern and southern Atlantic coast due to the regularity and relatively high wind speed. While the photovoltaic and the CPV installations, are/will be located in the eastern and southern regions. Mainly Tafilalet-Draa region, in which the projects Noor are/will be implemented and the sun is usually shining for the most of the year. Figure 9.2 shows the exciting and projected solar and wind large energy installations.

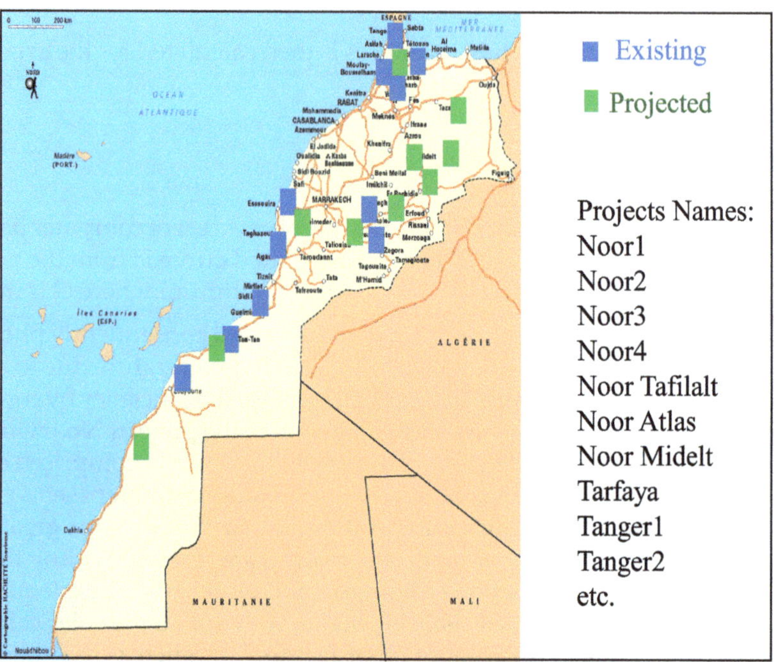

Figure 9.2: Solar and Wind Large Energy Projects. Existing (in blue) and projected (in green).

3.3 Towards a 100 per cent Electricity from Renewable Energies

During the COP21 (The 21st session of the UN Conference of the Parties, held in Paris, France, 2015), the Moroccan government increased the part of electricity from renewable energies to 52 per cent by 2030 (20 per cent solar, 20 per cent wind and 12 per cent hydro). In the COP22 (The 22nd session of the UN Conference of the Parties, held in Marrakech, Morocco from 7-18 November 2016) 48 countries committed to try to meet 100 per cent domestic renewable electricity production as rapidly as possible. Morocco, the host country of COP22, is among. They are united as the Climate Vulnerable Forum (CVF) [13]. During COP22, it was successfully demonstrated that the implementation of the Paris Agreement is underway and the spirit of multilateral cooperation on climate change continues.

3.4 Towards a 52 per cent Electricity from Renewable Energies by 2030

The important question we need to ask is what contribution can renewable energies make in the future to the electricity production?From the beginning of the 21st century, electricity demand in Morocco grew at 6.6 per cent a year with a structure increasingly resembling that of developed countries [19]. This increase last for about ten years, and was driven by the generalization of electricity access, improved living standards and demographic growth. It thus went from 15,540 GWh in 2002 to 33,523 GWh in 2014 [19]. From 2014 the electricity demand slowed to about 3 per cent increase and stabilized. Figure 9.3 shows the electricity demand and the part converted from renewables during the mentioned period above. The year 2015 was a turning point in the energy transition process with the introduction of the CPV Noor1 in the electricity market (the total electricity converted from renewables reached about 11.6 per cent).

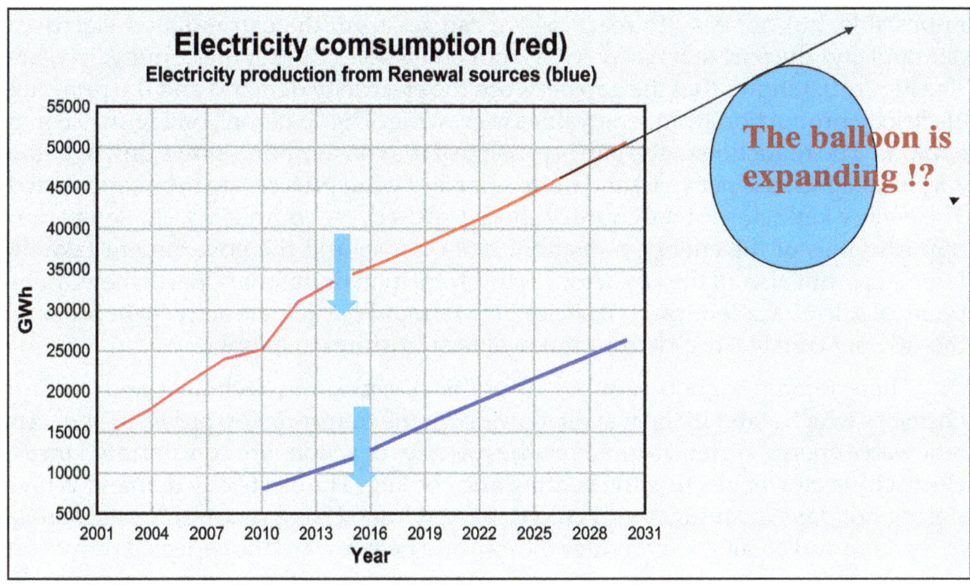

Figure 9.3: Morocco Electricity Demands 2002-2030 (Top curve) and Electricity from Renewable 2015-2030 (Bottom curve). The arrow shows the year of the turning point by the introduction of the CPV Noor1 in the grid.

In 2020 the renewal installations capacity will represent 32 per cent of the 7,992 MW, will reach 15,946 MW in 2020, 20,070 MW in 2025 and 24,800 MW by 2030. To reach the targets set for 2020 and 2030 Morocco will have to develop between 2017 and 2030 additional renewable energies electricity production capacity amounting to about 10,100 MW of which 4,560 solar, 4,200 wind and 1,330 MW hydro [11-13]. In Figure 9.3 we have extrapolated the electricity consumption for the period 2016-2030 assuming a stabilized increase of about 2.6 per cent per year (it is the projected increase in consumption during the coming 10 years). Also we have extrapolated the part of electricity converted from renewable in order to reach the 52 per cent target in 2030 by assuming that expected projects will be realized.

3.5 Is 100 per cent Renewable Energy Electricity Possible?

To ensure the energy transition towards a nation supplied by 100 per cent renewable energies, Morocco must radically modify its energy sector and take the right steps. The challenge is not so much a lack of energy resources but to fundamentally transform the way in which the energy system is structured. Renewable energies are based on decentralized production, have a horizontal supply chain and need an infrastructure and market which will permit the emergence and participation of new actors, including citizens. We want to note here, that in addition to the government's large implantations, there are growing number of homes and industries that are exploiting renewable energies (mainly solar), and their production is not included in our estimations.

To move towards a 100 per cent electricity from renewable energies is not impossible, but not easy to reach as we can see from the extrapolated electricity demand and the part to be converted from renewables during the coming 30 years. The Figure 9.3 shows that the gap between the electricity demand and the projected electricity production from renewables is growing. The "balloon" on Figure 9.3 may explode. The reduction of the gap is possible if the consumption slows down (below 2.6 per cent increase per year) and more solar and wind projects are to be constructed. The energy key actors of the country should address the complexity, challenges and opportunities of the energy transition. Not only should the government take the large steps, but also all the key actors in this transition (industrials, parliamentarians, political actors, academics, civil society, *etc.*) should participate actively in order for this 100 per cent electricity from renewables transition to happen.

There remain a good number of political, economic, technical and cultural obstacles which stand in the way or slow down the transition towards a 100 per cent renewable energy system. Moreover, the majority of actions are concentrated on the electricity sector, neglecting the heating and cooling. The electricity demand is high during hot days in summer and cold days in winter. Using renewable alternatives for heating and cooling may reduce the gap and help toward the expected transition.

4. Conclusions

The Moroccan renewable energies potential is tremendous due to its geographical situation. Solar radiation is intensive in most parts of Morocco and mainly in the east and south. We have wind potential in the coastal regions, specifically in the southern and northern regions (large and steady wind system). Hydraulic is limited but most of it is used and contribute to the electricity needs. To move towards a 100 per cent electricity from renewable energies is not impossible, but not easy to reach. The challenge for the country is not so much a lack of energy resources but to fundamentally transform the way in which the energy system is structured. This transition is possible if the consumption slows down (below 2.6 per cent increase per year), alternatives for heating and cooling are used and also with the involvement of the private sectors in solar and wind projects.

REFERENCES

1. McEvoy A.J. "Outlook for photovoltaic electricity. Endeavour, New Series, 17(1), 1993, pp.17-20
2. Odeh I., Yohanis Y.G, and Norton B.; Solar Energy, 2006, 80, pp.850-860
3. Mekkaoui Alaoui I. "Photovoltaic solar water pumping: A complement to usual irrigation systems in the Ziz Valley (East of Morocco)", International Conference on Renewable Energies (INCORE), Cairo, 3-6 February, 2016.
4. Zafar S., Renewable Energy, Solar Energy, January 24, 2016
5. NASA Langley Research Center Atmospheric Science Data Center; New *et al.*, 2002.
6. Schinke B., *et al.* (2016) « Middle East North Africa Sustainable Electricity Trajectories"
7. https://www.bicc.de/fileadmin/Dateien/pdf/press/2016/
8. Ettaik Z., «Les énergies renouvelables au Maroc: Bilan et Perspectives ».
9. Ministère de l'Énergie, des Mines, de l'Eau et de l'Environnement du Maroc, (2015)
10. Kousksou T., and al."Renewable energy potential and national policy directions for sustainable development in Morocco" Renewable and Sustainable Energy Reviews 47, 2015, pp 46-57.
11. MASEN (2016) "Noor Ouarzazate" http://noorouarzazate.com/3/
12. Algora C., Rey-Stolle I., Handbook on Concentrator Photovoltaic Technology, Wiley, May 2016.
13. El-Katiri L., 'Les atouts du Maroc dans le domaine de l'énergie verte' Policy Brief, December 2016, PB-16/33.
14. Ben Hayoun, M. (2016) "Transition énergétique: comment évoluera le mix électrique entre 2015 et 2030 » http://lematin.ma/journal/2016/
15. García I., and Leidreiter A., "A. Roadmap for 100 per cent Renewable Energy in Morocco", World Future Council (WFC), 2016, pp5-29
16. Mekkaoui Alaoui I., Doctorat de 3ieme Cycle, 'Etude Comparative des Modèles de caractérisation I-V des photopiles, Juillet 1984, USTL, Montpellier, France.
17. Charles J.-P., Mekkaoui Alaoui I., and Bordure G., "Etude Comparative des Modèles a Une et Deux Exponentielles en Vue d'une Simulation Précise des Photopiles.", Revue Phys. Appl. 19, (1984) pp. 851 857.
18. Charles J.-P., Mekkaoui Alaoui I., and Bordure G., "A critical Study of the Effectiveness of the Models for I V Characterization of Solar Cells."Solid State Electronics, 1985, 28(8), pp. 807-820.
19. Skoplaki E., Palyvos J.A., "On the temperature dependence of photovoltaic module electrical performance: A review of efficiency/power correlations", Solar Energy, Volume 83, Issue 5, May 2009, pp. 614–624.

20. Ortiz-Rivera E. I. and Peng F. Z., "Analytical Model for a Photovoltaic Module using the Electrical Characteristics provided by the Manufacturer Data Sheet", 2005 IEEE 36th Power Electronics Specialists Conference, 2005, pp.2087 - 2091

21. ONEE (Office Nationale d'Electricité), 'Facteurs de réussite d'une intégration des énergies renouvelables: expérience marocaine et perspectives', 2015, http://www.somelec.mr/IMG/pdf/energies_renouvlable

Chapter 10

Evaluating the Performance of the PV-Modules Available in Palestinian Market

Imad Khatib and Makawi Hraiz

Palestine Polytechnic University,
Hebron, Palestine
E–mail: mekawi@ppu.edu

The utilization of solar energy in Palestine has played a significant role during the last three decades as it saved around 15 per cent of conventional energy that might have been used in building sector. Figures of 2016 suggest that more than 70 per cent of the Palestinian households are using the solar domestic hot water systems for heating up water for their domestic uses. This figure is the highest in the region In addition to the thermal conversion of solar energy, Palestinians have recently realized the huge benefits of the photovoltaic direct energy conversion systems in particular with their economical feasibility. The Palestinian cabinet has positively responded by putting enabling policies and regulations; including the feed-in-tariff and net-metering.

Today, there exists around sixty companies working directly on importing and installing PV-Systems of different capacities and in response to the increasing demand and for ensuring the quality of the imported PV-Modules, the Polytechnic University has established a unit for testing the performance of PV-Modules in the framework of a World Bank funded project and in cooperation with the Florida Solar Energy Centre (FSEC) based on an indoor flash test bed simulator (FTBS). This unit complements the previously established unit for testing the thermal performance of solar hot water systems (collectors and the thermal storage). Both testing units are ISO17025 certified. The indoor testing unit enables conducting performance testing of the PV-Panels using Standards Test Conditions (STC) and other variable testing conditions that simulate PV-Modules' possible working environment.

Keywords: *Photovoltaic, Flash test bed simulator, PV efficiency.*

1. Introduction

Countries of potential solar radiation for energy conversion systems have been able to lower their carbon foot print by leveraging from the abundant renewable energy. Located on the solar belt, Palestine has long benefited from solar energy, which is on daily average on annual basis exceeds 5.2 kWh/m -day, by converting it into thermal energy for heating domestic water. With the advancement in direct energy conversion system components; PV modules, inverters and storing batteries including together with the reduction to their costs, the utilization of solar energy in power generation turned to be of paramount important to the Palestinians. Palestinians are still dependent on imported power as they could only generate around 10 per cent and the rest are provide mainly from Israeli Electric Power Company through connecting power networks that feed most of Palestinian areas. Several years ago, Palestinians succeeded, with technical assistance provided by humanitarian, to providing electric power to remotely marginalized rural areas mainly to power community clinic or a shared refrigerator [1]. During the last two years the Palestinian Authority has embarked on preparing policies and guidelines to encourage the use of PV power systems; mainly grid-tied, to mitigate the impact of energy dependency. The National Energy Efficiency Action Plan (NEEAP) suggests that by the 2020, Palestinian Authority could produce around 10 per cent of its power from solar energy [2]. Such encouragement policy requires that proper efficient system components to installed and to protect increasing customers from un-efficient PV components commercialized in the market.

The work presented in this paper is cooperation among Palestine Polytechnic University (PPU), Palestinian Energy and Natural Resources Authority (PENRA) and the Palestinian Standards Institute (PSI) aiming at providing technical information about the PV modules mostly used in local market to conforming their declared performance to the performance tests carried out in PPU Solar Energy Testing Laboratory (PPU-SETL). PPU-SETL is equipped with a flash test bed simulator that could simulate the different working environment in addition to producing the standard test conditions (STC) required to test PV-Modules; *i.e.* Irradiance of 1000 W/m^2 and a ambient temperature of 25 °**C**.

By analyzing the results of the different tests a guidelines will be produced in cooperation with the PENRA and PSI that will serve the stakeholders.

In addition, investigation on different techniques that may be used to enhance the performance of the PV-Modules will be carried out.

It is worth mentioning that the FTBS tests can produce the following electrical performance characteristics:

- ☆ Current-voltage curve.
- ☆ Power-voltage curve.
- ☆ Open circuit voltage.
- ☆ Short circuit current.
- ☆ The peak power Pm.
- ☆ The maximum power point MPP voltage Vm and current lm.
- ☆ Fill Factor FF.
- ☆ The module efficiency η.

☆ The test temperature T.

By interpretations of analysis, the aim guidelines will be produced.

2. Testing the Performance of the PV Module

The PV module efficiency determines its power output per its area. It is well documented that the maximum efficiency of the PV module depends on module's physical characteristics, which is usually declared as specifications by the manufacturer. The PV module used in the testing facilities is a SUNTECH 140 Wp. In Table 10.1, module specifications at STC are listed in accordance to it's label.

Table 10.1: Specifications of PV Module (Manufactured by SUNTECH)

Rated Maximum Power (Pmax)	140W
Tolerance(Tol)	0 - □3 per cent
Open- circuit voltage (Voc)	21.85V
Short-circuit current(Isc)	8.59A
Fill Factor (FF)	0.748
Cell Technology	Poly-Si

Both modules where installed simultaneously in two different locations, in Ramallah and in Jericho in two 5kWp photovoltaic power generation systems. The two cities are 24 kilometers apart and, while Ramallah is located at an altitude of 800 ASL at 31.89 latitude and 35.2 longitude, Jericho is located at the lowest point on earth of +250 BSL and at a latitude of 31.86 and a longitude of 35.45. Ramallah climate is strongly influenced by Mediterranean circulation while Jericho is strongly influenced by the high pressure developed at the rift.

PV efficiency could be calculated directly using (Eq.1):

$$\eta_{max}=P_{max}/[Ed*Ac] \quad (1)$$

where,

η_{max} Efficiency of PV module, per cent

P_{max} Maximum Output power, W

Ed Radiation at STC, $1000W/m^2$

Ac Area of PV module, m^2

Relying on the module label, an efficiency of 14.1 per cent is considered, which is considered low based on NREL[1] (Figure 10.1).

Testing the PV module took place indoor using the FTBS of PPU and in accordance to ASTM E 1036 standards with purpose to determine the rated power (P_{mp}) at STC along with other performance parameters; V_{oc}, I_{sc}, V_{mp} and I_{mp}. It should be pointed out that only one module from same brand was tested.

1 https://cleantechnica.com/2014/02/02/which-solar-panels-most-efficient/

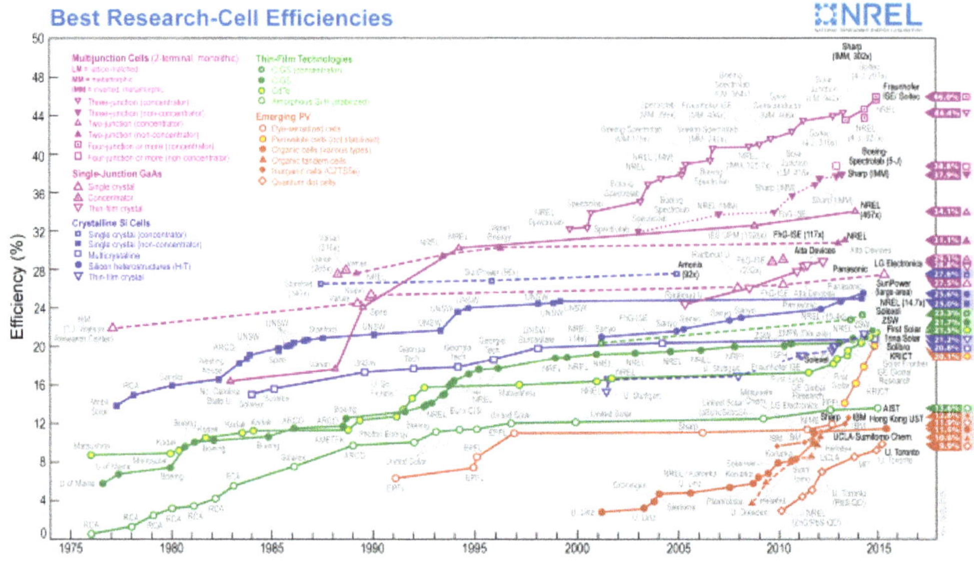

Figure 10.1: NREL Ranges of different Modules Efficiencies.

The test report on the module performance usually includes:

- ☆ Module identification and description,
- ☆ Irradiance identification,
- ☆ Performance parameters highlighted above,
- ☆ I-V curve with intercepting values intercepting I_{sc} and V_{oc} and location of P_{mp}.

The possibility of controlling input conditions indoor facilitated testing then using different set conditions in order to produce relevant coefficients of the module. Two sets of performance testing were used. First, an STC irradiance value was kept constant at 1000 W/m^2 and three ranges of ambient temperature were used; 15, 25 (STC) and 35 °C, thus allowing to simulate the average difference between the seasonal temperatures of Jerusalem and Jericho areas which is in the range of 8 - 12 °C (Figure 10.2).

In the second sets work with the STC ambient temperature with four different ranges of irradiances, including the highest at STC. This was performed to simulate the Jericho unique climate where higher pressure compared to that at sea play a significant role in Jericho's ambient temperatures.

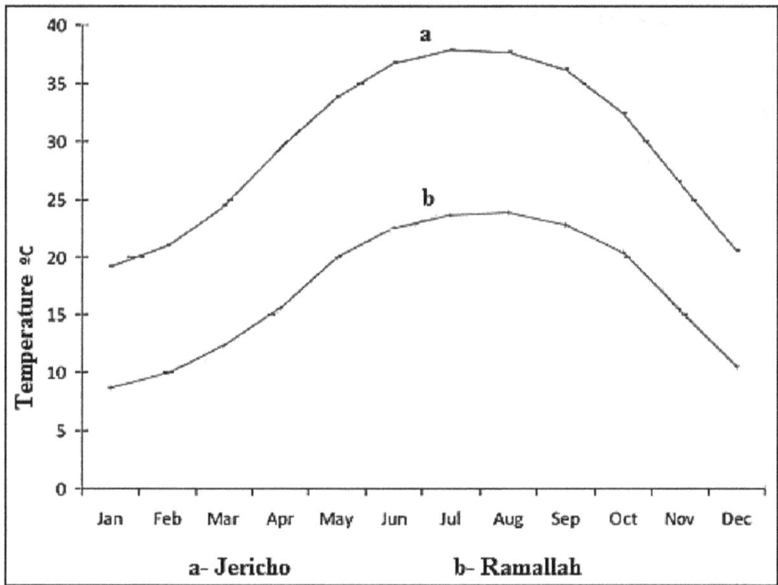

Figure 10.2: Mean Temperature Variability between Jericho and Ramallah in 2010 [4].

3. Experimental Results and Analysis

3.1 Influence of the change in ambient temperature on I -V curves

Obviously, temperature can vary quite a bit depending on location and season. The solar PV cell temperature rises due to the rise in ambient temperature, the main effect is to reduce the voltage and power output available at most currents. Same for PV module, where open circuit voltage rapidly decreases with increases in ambient temperature [2]. The resulted module output will therefore reduce by a fraction that depends on the temperature coefficient (C_v) which varies depending on the cell's materials. Modified output could be calculated based on equation:

$$P_{max} = I_{mp(rated)} \times V_{mp(modified)} = I_{mp} [V_{mp(rated)} \times (C_v \times (T_{amb} - 25))] \quad (2)$$

Usually, C_v is a negative V/°C ratio which is much decisive in modifying operating voltage but could slightly affects operating currents. [3].

Testing using the three ranges of temperature produced same short-circuit current, however effects on open circuit voltage is clearly noticed (see Table 10.2).

Based on the results obtained a suggested average temperature coefficient is computed -0.04 V/°C. It should be noted her that the -0.55 V deviation of the module labeled V_{oc} and measured at STC V_{oc} was not considered and calculations were made using the resulted performance outputs. I - V characteristic curves for the set of ambient temperatures could be seen in Figure 10.3.

Table 10.2: Module Performance at different Controlled T_{amb}

Temperature	15	25	35
Isc(A)	8.575	8.576	8.57
Voc(V)	22.7	22.422	21.9
Pm(W)	138.2	136.28	132.6
Ipm(A)	7.71	7.7829	7.8
Vpm(V)	17.91	17.51	17
η (per cent)	13.2	12.92	12.6
F.F	71.01	70.87	70.59

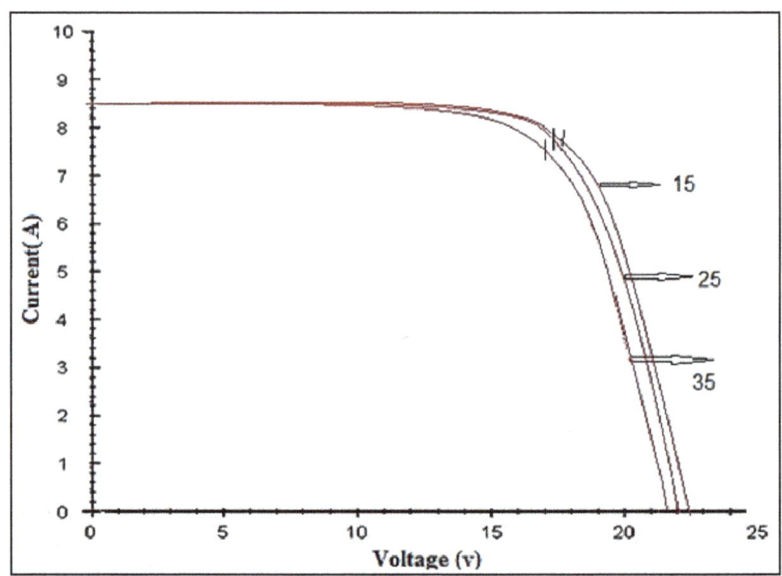

Figure 10.3: Measured Current-Voltage Characteristic Curves under Various Module Temperatures at 1000 W/m².

Curves show that outputted V_{oc} drops by about 0.04 V for every degree Celsius increase in ambient temperature.

In Figure 10.4, open circuit voltage as a function of ambient temperature could fit the model equation suggested by Musembi *et al.* (2013) and described elsewhere and could be further used to in a linear extrapolation way to give the amount of activation energy value (E_a) at zero T_{amb} which computed at 0.76 eV.

3.2 Module Performance at different Controlled Irradiances

Experimental investigation was further carried out simulating a range of irradiance with minimum value of 400 W/m^2 increasing steadily to 600, 800 and to STC of 1000 W/m . In all controlled values of irradiance, temperature kept constant at STC of 25 C.

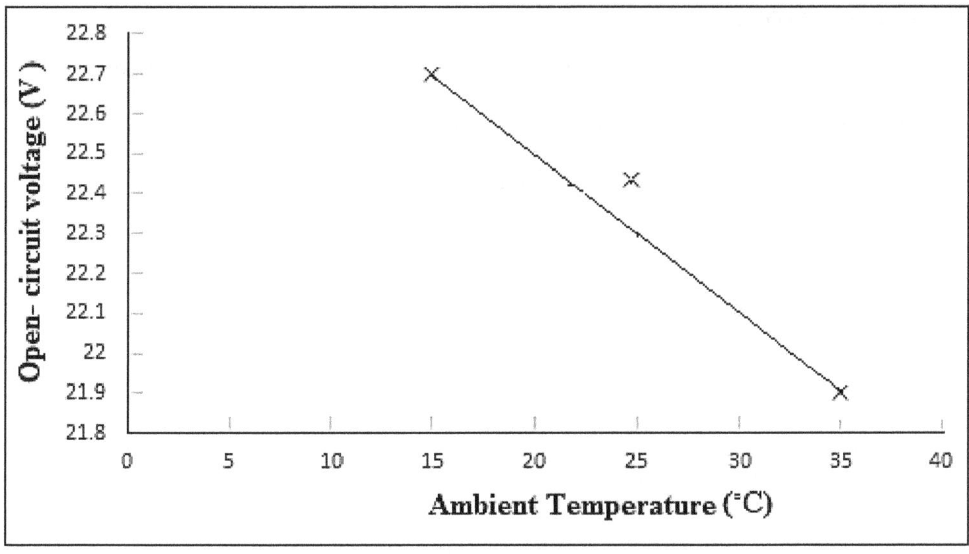

Figure 10.4: Measured Open Circuit Voltage under Various Module Temperatures.

Results obtained have showed the effect of on the short circuit current at each level of incident radiation. A linear representation of the generated short circuit current with the different levels irradiances illustrates the shading effect on the module performance as well (Figure 10.5).

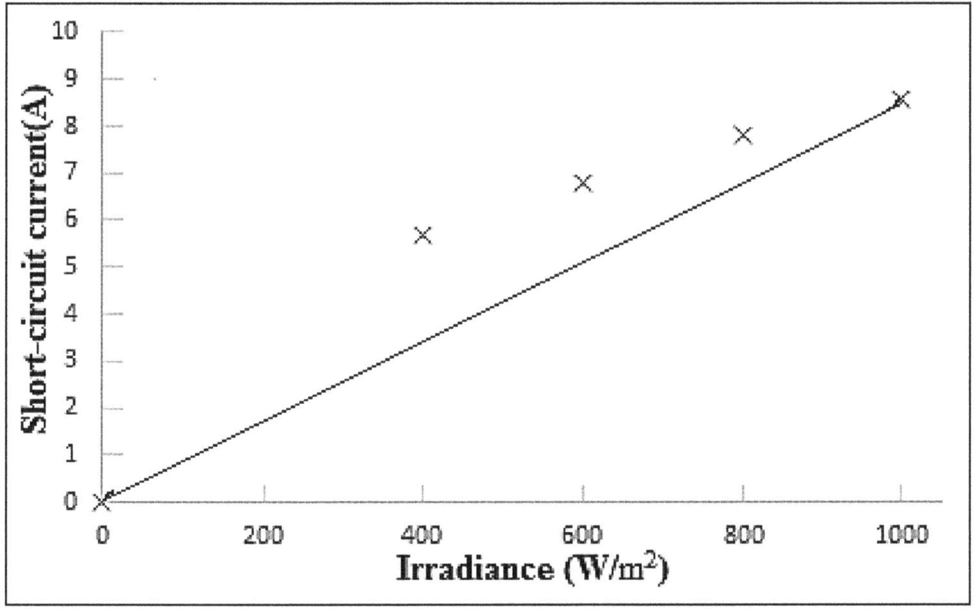

Figure 10.5: Measured Short Circuit Current under Various Irradiance.

4. Conclusions

Tests carried out for the PV module showed the clear effects of the variation in Temperature and Irradiance on the performance of the PV systems. Further work is going to focus on how too technically reduce adverse decrease in module efficiency.

REFERENCES

1. Marwan M. Mahmoud, ImadIbrik, Field experience on solar electric power systems and their potential in Palestine,Renewable and FIG Sustainable Energy Reviews, 2003;7:531-541.
2. M. Zdravkovic´, A. Vasic´, C´. Dolic´anin, K. Stankovic´, P. Osmokrovic, Temperature Effects on Photovoltaic Components Characteristics, APPL. MATH. INFORM. AND MECH. vol. 1, 1 (2009), 29-36.
3. Swapnil Dubey, Jatin Narotam Sarvaiya, Bharath Seshadri, Temperature Dependent Photovoltaic (PV) Efficiency and Its Effect on PV Production in the World,Energy Procedia 33 (311 – 321),2013.
4. Matthew Richard Jad Issac,Analysis of climate variability and its environmental impacts across the occupied Palestine territory,2013.

Chapter 11

Updates on the Transforming Commitments into Action-DOST MIMAROPA Experiences

Bernardo T. Caringal

Provincial Science and Technology Director, Marinduque, DOST MIMAROPA, 4th Floor, PTRI Building, Bicutan, Taguig City, Philippines
E-mail: dostmar@yahoo.com

The Philippines is an archipelago of 7,107 islands situated in mainland Asia and separated from it by the West Philippine Sea. The total land area is approximately 300,000 sq km (115,831 sq mi).

It has 17regions and one of the newest which was established in 2003 is the MIMAROPA which comprises of the island provinces of Mindoro Occidental, Mindoro Oriental, Marinduque, Romblon and Palawan.

The Philippines is located in the Northern-side of the globe. It is between the equator and the Tropic of Cancer. Thus, being close to the equator has an advantage by the angle of the solar energy reaching the surface of the earth. This is exactly the reason why Philippines experiences hot climate and dry season.

As reported by DOST-PAGASA (Philippine Atmospheric, Geophysical, and Astronomical Services Administration), the country is visited by an average of 20 typhoons each year. And with the eminent effect of climate change, the typhoons are getting stronger. One of the latest is typhoon NINA internationally known as Nock-ten, which devastated the MIMAROPA especially Marinduque, Mindoro and Romblon last Christmas, December 25, 2016. It left the provinces with destroyed houses, downed trees and lampposts, blocked highways and widespread power interruptions.

According to National Disaster Risk Reduction and Management Council (NDRRMC), the province of Marinduque and two (2) towns of Romblon have experienced power interruption due to typhoon Nina. It took at least a month before the Marinduque Electric Cooperative (MARELCO) restored the power supply in the province. People struggled without electricity and battled every night in dim light.

The Philippines is abundant in sunlight and DOST-MIMAROPA strategized on how to make use of it. With these, the Department of Science and Technology-Marinduque introduced the use of Solar Energy System (SES) in the province with the support of DOST-MIMAROPA regional office and different Local Government Units who provided budget for the said project. Through the SES, residents of these areas had their mobile phones charged allowing them to keep in touch with their relatives. Most importantly, rechargeable lights and flash lights were once again recharged ensuring that they will be spending the night with stable supply of power. In the process, converted electrical power from the sun became the midnight oil that gave them light till dawn.

Way back in the past years, this project is just a plan. Year 2014, the plans for this project were presented in India organized by Centre for Science and Technology of the Non-Aligned and other Developing Countries (NAM S&T Centre). The said presentation contains the plans and proposed budget allotment for the utilization of available renewable energy in the Philippines. The plan for nationwide adoption of Solar Energy System for DOST-assisted projects was initiated by DOST-MIMAROPA who saw the potential to provide uninterrupted electric supply in the region. Thus, Marinduque Provincial Science and Technology Center started the advocacy through series of seminars and technology demonstration in the province and across the region. As a result, the SES project is continuously growing and transforming lives of people up to present. But looking now, it is fully implemented and proved to be beneficial for the people in the community. National Government Agencies like Department of Health-MIMAROPA, and Department of Social Welfare Development-MIMAROPA and Department of Education is consistently supporting the solar energy system as they can see its benefits and advantages to communities especially in the Geographically Isolated and Disadvantage Areas (GIDA).

At present, DOST-Marinduque installed SES for the 3 Rural Health Units in the province, 22 out of 37 Barangay Halls in Mogpog and 7 secondary schools as evacuation center and small scale SES for households under CEST, 1 Community Empowerment through Science and Technology (CEST) Area and 3 processing centers. These SES installation province wide is a success and the establishment of the solar panel is the result of the advocacy of DOST-MIMAROPA to provide meaningful area improvement in rural communities across the region through innovative solutions.

Moreover, the impact of said technology was also realized through adoption in Island Municipality of Lubang in Occidental Mindoro and a Region-based SES project for selected RHU's (with RxBox) in the 5 provinces of MIMAROPA. This innovation brought major contributions for the awards received by PSTD Bernardo T. Caringal as 1st runner-up for the Search for Outstanding Provincial Science and Technology Director for the year 2015. Same project was highlighted for the nomination of Regional Director Ma. Josefina P. Abilay of DOST-MIMAROPA, in which in turn voted her as finalist for the BCY Foundation Innovations Award held last February 20, 2017.

Considering its history of implementation, from a Province, going to Regional, we are planning to make it into National level. DOST-MIMAROPA's objective is to widen and strengthen the project for its future usage.

1. Discussion

Facts and Figures about the Philippines

The Philippines is a Southeast Asian country in the Western Pacific, an archipelago consisting of 7,107 islands situated in the mainland Asia. The total land area is approximately 300,000 sq. km. (115,831 sq mi), categorized broadly under three main geographical divisions: Luzon, Visayas, and Mindanao.

The country is blessed with abundant natural resources and some of the world's greatest biodiversity.

The Philippines' location on the Pacific Ring of Fire and close to the equator, however, makes it prone to earthquakes and typhoons (an average of 20 typhoons or tropical storms each year). Cyclones develop over warm seas near the Equator. Cyclones begin in tropical regions, such as northern Australia, South-East Asia and many Pacific islands. In fact, in 2013, the country was struck by the Super Typhoon "Yolanda" (with International name "Haiyan") and was considered as one of the strongest typhoon that lashes the Philippines leaving 7,200 people dead or missing and Billion-worth of damages to Agriculture and infrastructures.

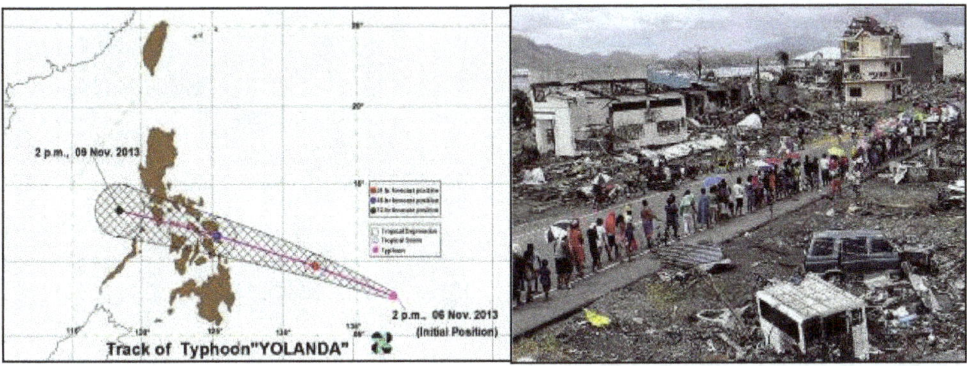

Figure 11.1: *Right*–Track of Typhoon "Yolanda" that devastated the Visayan Regions of the Philippines. *Left*–People of Tacloban, Leyte fell in line for relief goods after the typhoon "Yolanda.

It has 17 regions and one of the newest which was established in 2003 is the MIMAROPA which comprises of the island provinces of Mindoro Occidental, Mindoro Oriental, Marinduque, Romblon and Palawan.

Facts and Figures about Region IV-B (MIMAROPA)

The South-western Tagalog Region, designated as MIMAROPA Region, is an administrative region in the Philippines. It was also formerly designated as Region IV-B until 2016. It is one of five regions in the country having no land border with another region (the others being Western Visayas, Negros Island Region, Central Visayas, and Eastern Visayas). The name is an acronym combination of its constituent provinces: Mindoro (divided into Occidental Mindoro and Oriental Mindoro), Marinduque, Romblon and Palawan. MIMAROPA comprises 5 provinces, 1 highly urbanized city (Puerto Princesa), 1 component city (Calapan), 71 municipalities 1,458 barangays and the total population of 2,559,791.

It is only last December 2016 that the region was heavily devastated by another, yet strong typhoon, the super typhoon "Nina" leaving at least P83.4 million damage to infrastructures in Marinduque, Mindoro and Romblon and displaced at least 132,908 families or 602,770 people from 785 barangays in affected provinces, forcing them to temporarily seek shelter in 145 schools.

Figure 11.2: Actual Footages of the Aftermath of Typhoon "Nina" Leaving at least One Month of No Electric Supply. The SES proved to be functional and beneficial to evacuees and the community in Bry. Yook, Buenavista, Marinduque.

Department of Science and Technology (DOST)

The Department of Science and Technology (DOST) is the premiere science and technology body in the country charged with the twin mandate of providing central direction, leadership and coordination of all scientific and technological activities, and of formulating policies, programs and projects to support national development.

In 1958, Congress passes a law establishing the National Science Development Board (NSDB) upon the recommendation of Dr. Frank Co Tui, who was tasked to survey the state of Philippine S&T during the Garcia administration. It was then revamped in 1982 as the National Science and Technology Authority (NSTA) and accorded broader policy-making and program implementing functions. It was then elevated in 1987 to Cabinet level and becomes the DOST in response to increasing demands for S&T intervention in national development.

The DOST MIMAROPA

MIMAROPA is the youngest among the 17 administrative regions of the country. It was created as a separate region through Executive Order 103 issued on May 17, 2003 to accelerate the social and economic development and improve the delivery of public services.

The acronym MIMAROPA stands for the provinces of the region, namely, Mindoro Occidental, Mindoro Oriental, Marinduque, Romblon and Palawan. There are two cities in this region, Calapan City in Oriental Mindoro and Puerto Princesa in Palawan. Among the region's provinces, Palawan is the biggest, accounting for almost 54.7 per cent of the regions' total land area while Marinduque is the smallest with an area of only 944.7 square kilometres or barely 3.4 per cent of the total land area of the region.

MIMAROPA's strongest potentials are on agriculture and ecotourism. Forty per cent of the Gross Regional Domestic Product comes from Agriculture and Fishery. It also intensifies food production not just for the region but for the CALABARZON and Metro Manila areas as well.

In September 2007, DOST Secretary Estrella F. Alabastro signed a Memorandum separating the management of DOST-MIMAROPA from DOST-IV with a common administrative personnel sharing with DOST-CALABARZON. DOST-MIMAROPA holds office at DOST Complex in Bicutan, Taguig City with its first Regional Director, Dr. Ma. Josefina P. Abilay.

DOST MIMAROPA Banner Program

Small Enterprise Technology Upgrading Program (SETUP) was launched in response to the President's call for more focused programs of assistance for Micro, Small, and Medium Enterprises (MSMEs). It is a nationwide strategy to encourage and assist MSMEs to adopt technological innovations to improve their operations and thus boost their productivity and competitiveness. The program enables firm to address their technical problems through technology transfer and technological interventions to improve productivity through better product quality, human resources development, cost minimization and waste management and other operations related activities. The program focused on 5 industry sectors, namely: Food processing, furniture, gift-decors-housewares (GDH)S, aquatic resources and horticulture and metals and engineering.

The DOST MIMAROPA also provided assistance to State Colleges/Universities, Local Government Units and organized groups through the DOST Grant-in Aid (DOST-GIA). Beneficiaries are provided with financial grants to S&T programs and projects to spur and attain economic growth and development by harnessing the country's scientific and technological capabilities.

In 2016 the DOST MIMAROPA assisted 42 and 20 projects under SETUP and DOST-GIA, respectively. About 50 per cent of that is categorized under food processing sector. Although those projects are mostly micro enterprises, it generated about 180 employments per month.

Mitigating the Effects of Climate Change through Solar Energy System

The MIMAROPA Region–comprised of the island-provinces of Occidental Mindoro, Oriental Mindoro, Marinduque, Romblon, and Palawan–while it is geographically and politically close to mainland Luzon, still has a long way to go in terms of keeping up with the social and economic growth of its nearest neighbors: Region IV-A, otherwise known as CALABARZON, and, of course, the National Capital Region.

The unreliability of the MIMAROPA Region's energy source for electric power has been, for a long time, one of the impeding factors that further exacerbate the present challenges of the region in the development of its agriculture, business and industry, health, and education.

The region's island-provinces are not connected to the main electric grid in Luzon. As such, and according to the Regional Development Plan 2011-2016, "The power generation for the region is through the Missionary Electrification of the Small Power Utilities Group (SPUG) of the National Power Corporation (NPC). The SPUG is responsible for providing electricity in areas where no private sector entity is willing or able to provide the same service at reasonable cost." Moreover, the RDP 2011-2016 also stated, "The energization of the region is through the eight electric cooperatives servicing the MIMAROPA provinces and municipalities." The eight electric cooperatives are: (1) Lubang Electric Cooperative, Inc. (LUBELCO); (2) Occidental Mindoro Electric Cooperative, Inc. (OMECO); (3) Oriental Mindoro Electric Cooperative, Inc. (ORMECO); (4) Marinduque Electric Cooperative, Inc. (MARELCO); (5) Tablas Electric Cooperative, Inc. (TIELCO); (6) Romblon Electric Cooperative, Inc. (ROMELCO); (7) Busuanga Electric Cooperative, Inc. (BISELCO); and (8) Palawan Electric Cooperative, Inc. (PALECO).

The National Electrification Administration (NEA), nonetheless, in its Electric Cooperative Fact Sheet as of September 2016, reported a 100 per cent accomplishment in the 1,434 total number of barangays energized across the region. Despite achieving 100 per cent energization, however, power interruptions or outages in the provinces have still been generally quite frequent, affecting not just the everyday lives of the people and the business environment in region, but also hampering the effective delivery of government services.

The latest Regional Economic Situationer for the 1st to 3rd Quarters of 2016 from the National Economic Development Authority MIMAROPA (NEDA-MIMAROPA) reported a total of 6,797 power interruptions with a total duration of 11,834.90 hours, or an average of 481 power interruptions with a total duration of 840.62 hours per quarter, across the region.

There are a total of 31 power plants/power barges under the eight electric cooperatives servicing the region's provinces. However, 12 of these do not operate 'round-the-clock, according to the National Power Corporation's (NPC) SPUG Power Plants/Power Barges Operational Report for Existing Areas report as of November 25, 2016. These 12 power plants/power barges in the region with limited operating times are the following: (1) Maniwaya, Mongpong, and Polo power plants, all located in the municipality of Sta. Cruz in the province of Marinduque, and all of which have eight-hour operations only; (2) The Cabra power plant, located on Cabra Island in the municipality of Lubang in the province of Occidental Mindoro, which has eight-hour operations only; (3) The power plants in the province of Romblon, particulary in Banton and Concepcion, and San Jose and Corcuera, which have eight-hour and 16-hour operations, respectively; and (4) The power plants in Cagayancillo, Linapacan, Agutaya and Balabac in the province of Palawan, which have eight-hour, 12-hour, and 16-hour operations, respectively.

DOST-MIMAROPA's Solar Energy System Project Plans Turned into Action

While the MIMAROPA Region is geographically and politically close to mainland Luzon, it still has a long way to go in terms of keeping up with the social and economic growth of its nearest neighbors, the CALABARZON Region and, of course, the National Capital Region. Energy sources for electrical power–requisite for a quality life in these times–is not that reliable. Electric power outages are generally quite frequent, affecting not just the everyday life of the people in MIMAROPA, but it also hampers the effective delivery of government services.

On top of all the other initiatives of the DOST that the Regional Office had to implement, particularly on enterprise development and community empowerment–harnessing available renewable energy such the Solar Energy System was initially explored through the attendance of Provincial S&T Director Bernardo T. Caringal from the province of Marinduque to a training on Renewable Energy and Energy Efficiency in New Delhi, India in 2012. With the knowledge gathered from the training, a series of forums on solar energy and demonstrations of the solar energy system were conducted in Marinduque in cooperation with the Marinduque State College School of Industrial Technology. These forums were aimed at encouraging municipal and barangay officials to push for budget allocation for the use of solar energy in their respective offices

As a result, continued initiatives (advocacy) for the adoption of SES in different evacuation centers (Barangay hall and DepEd Schools), as well as for the power requirements in the Rural Health units in the province of Marinduque were implemented. Rural health units (RHUs) and their extensions, barangay health stations, are vital services in a community that should always have a reliable power source. Disruptions in basic health services, treatment, and even spoiling of vaccines due to power outages, can mean severe harm, or even death to patients.

It all started with DOST-MIMAROPA's initiative of installing a solar energy system in the barangay health center in Brgy. Dolores. Both DOST-MIMAROPA and DOH-MIMAROPA saw how the project changed the lives of the locals in the said barangay. The technology supplied the barangay health center with a reliable power source for the health center's lighting fixtures and ventilation, which, in turn, helped with the continuous provision of health services by its doctors, as well as with the comfortable and safe birth giving of mothers.

The year after, through the initiative again of DOST-MIMAROPA, both the agency and DOH-MIMAROPA agreed to install solar energy systems for RxBoxes–telehealth devices used for medical check-ups–throughout Marinduque, starting with the Rural Health Unit (RHU) in Brgy. Bantay in Boac. The project was worth PhP 200,000.00. With the installation of the solar energy system, patients can rest assured medical check-ups using the RxBox technology can be carried out effectively and efficiently even during electric power outages. Furthermore, the said RHU also serves as an evacuation center during natural calamities, and the reliable source of power provided by the solar energy system meant that appliances such as electric fans, lighting fixtures, and communication devices keep functioning even if electric power lines in other areas are down.

Just recently, the DOST-MIMAROPA have entered again into another project with DOH-MIMAROPA to install solar energy systems in 29 RHUs spread across the region, starting this year, 2017.

Table 11.1: Areas Installed with Solar Energy System

Nature	***Location***	***Funding Agency/ies***	***Amount***
Rural Health Unit			
Rural Health Unit - Dolores	Brgy. Dolores, Sta. Cruz, Marinduque	DOH-MIMAROPA DOST-MIMAROPA	P30,000.00
Rural Health Unit - Boac	Brgy. Bantay, Boac Marinduque	DOH-MIMAROPA DOST-MIMAROPA	P230,000.00
Rural Health Unit - Maniwaya	Brgy. Maniwaya, Sta. Cruz, Marinduque	DOH-MIMAROPA DOST-MIMAROPA	P250,000.00
Solar Energy System for Rx Box			
Region-wide RHUs with Rx- Box	29 RHUs	DOH-MIMAROPA DOST-MIMAROPA	P3,400,000.00
Evacuation Centers			

Nature	Location	Funding Agency/ies	Amount
Barangay Halls in Mogpog, Marinduque	Argao Banto Balanacan Capayang Dulong Bayan Gitnang Bayan Guisian Hinadharan Lamesa Laon Malayak	LGU-Mogpog DOST-MIMAROPA	P6,000,000.00
	Market Site Paye Sayao Sumangga Villa Mendez		
Secondary School as alternative evacuation center	Mogpog Central (Distict office) Mogpog National Comprehensive High School Butansapa NHS Puting Buhangin NHS Sayao NHS Argao NHS Balanacan NHS	LGU-Mogpog DOST-MIMAROPA	P1,000,000.00
CEST area	Binunga Elementary School	LGU-Buenavista DOST-MIMAROPA	P250,000.00
Processing Centers (DOST assisted Projects)			
Pilot Testing of Locally Designed Vacuum Fryer for Marinde Products	Brgy. Ino, Mogpog, Marinduque	LGU-Mogpog DOST-MIMAROPA	P150,000.00 (for SES alone)
Malunggay Crackers Production Facility		LGU-Mogpog DOST-MIMAROPA	P75,000.00 (for SES alone)
Coconut Processing Facility	Brgy. Ino, Mogpog, Marinduque	LGU-Mogpog DOST-MIMAROPA	P75,000.00 (for SES alone)

Breaking through Technology to Counter the Effects of Typhoon

Solar energy system of DOST-MIMAROPA has given a great help to people in many ways, especially during the aftermath of typhoon Nina last December 25, 2016 which shattered the province of Marinduque and Romblon resulting to power interruption.

According to National Disaster Risk Reduction and Management Council (NDRRMC) a total of 8 areas in MIMAROPA; Boac, Marinduque, Buenavista, Marinduque, Gasan, Marinduque, Mogpog, Marinduque, Santa Cruz, Marinduque, Torrijos, Marinduque, Looc, Romblon and Banton, Romblon have experienced power interruption due to typhoon Nina. For weeks since then, people residing in these areas strived with what was left of their resources. It took at least a month before the Marinduque Electric Cooperative (MARELCO) restored the power supply in the province. People struggled without electricity and battled every night in dim light.

With the *barangays* and schools serving as evacuation centers equipped with solar energy systems, evacuees managed to keep their communication lines open as they could charge their mobile devices as well as transmittal of reports (by teachers) through the use of their computer and printers connected to the SES.

DOST-MIMAROPA Target for the Coming Years

With the vision to widen and strengthen the Solar Energy System and further introduced it to National level, DOST-MIMAROPA will continue to provide funds with the strong linkage from different National Government Agencies (NGAs) to continuously strengthen the promotion of SES to make the Region powered by Solar Energy System. Through the achievement of SES in the succeeding years we will able to lower cost on electrical bills in different region and to the country as a whole.

Improved processing equipment which are in demand by the catered MSMEs in the manufacturing sector of the DOST-MIMAROPA will be also tapped. The potential of a commercialized solar-powered ice plant cum storage facility will be also explored to cater the need of the island provinces of the region for sustainable aquatic and marine resources.

REFERENCES

1. https://www.google.com.ph/#q=typhoons+hit+philippines+a+year and *
2. https://en.wikipedia.org/wiki/Mimaropa
3. http://secure.gnarlysunset.com/asia/philippines/mimaropa
4. https://sussle.org/t/Mimaropa
5. http://www.rappler.com/move-ph/issues/disasters/157020-oriental-mindoro-typhoon-nina-catastrophe

Chapter 12

An Overview of Solar Energy Landscape, Solar Radiation and Solar Cells Studies in Limpopo Province of South Africa

T.S. Mulaudzi and N.E. Maluta

Department of Physics,
University of Venda, P/Bag X5050,
Thohoyandou 0950, South Africa
E-mail: Sophie.mulaudzi@univen.ac.za

South Africa is well endowed with solar energy resource potentials and it has become an economically viable and environmentally preferable alternative to fossil fuels. In spite of the fact that some significant measurements of solar radiation were made in some developed and developing countries, this information is not available for most of the areas, especially in Africa. Due to lack of grid lines in the remote rural areas in South Africa, the use of the solar energy becomes more important and the government is introducing a lot of initiatives. These initiatives are spread throughout the whole SA. Some PV solar plants are developed in Limpopo Province such as Soutpan Solar Park which produces 28 MW and Tom Burke Solar Park which is located at Lephalale. Due to high electricity tariffs, Vuwani Science Resource Centre has built a small power producing plant. It is of this view that this study is done to assess the development of the renewable solar resources and its impact to the society. An attempted has been made to study the performance of solar cells and panels under outdoor conditions to evaluate the suitability of different solar cells/panels. In addition, we also conducted some studies on dye-sensitized solar cells for efficiency studies.

***Keywords**: Solar energy, Dye-synthesized solar cells, Efficiency, Solar park, Photovoltaic panels.*

1. Introduction

The global concern about energy crisis and man-made climate change is driving countries to use clean energy technologies. South Africa (SA) has committed to take mitigating action that would reduce South Africa's emissions by 34 per cent below the Business As Usual (BAU) trajectory by 2020. The decline scenarios for carbon emissions were taken into account for the development of the Integrated Resource Plan (IRP) 2010-2030 [State of Renewable Energy in South Africa 2015]. Lack of energy even in the form of electricity hinders the country's economic development. With the population growth continuing to outstrip electrification, just in Limpopo Province, 2011 and 2016 households were 1 418 102 and 1 601 083 respectively this indicates that the number of households is increasing, so the projection of enormous growing of households without energy access is a reality. Income spent security electricity has shown to be disproportionate, hence in most rural areas, the families are still collecting firewood which has detrimental impacts on the environment. According to Census 2016, some of the challenges as perceived by the household in Limpopo are the cost of electricity and lack or inadequate employment opportunities. The tons of coal are dwindling [www.statssa.gov.za]. The Tshikondeni coal mining that was situated in Limpopo had shut down, this also confirms the reduction of the amount of coal in the country. Also, due to the fact that coal is becoming a pariah fuel of the world, the department of energy in South Africa (SA) has committed to increase the renewable generation of energy. Bearing in mind that coal usage upsurges the amount of Carbon Dioxide (CO_2) in the air and hence the depletion of ozone layer in the atmosphere. Besides environmental benefits, solar energy provides electricity to remote areas where grid electricity is economically unfeasible.

The focus of this work is primarily based on an overview of the developments made up to this far on the landscape of solar energy in Limpopo Province. The application of renewable energy technologies gives the change of systems. Installation of solar energy devices such as solar water heaters, photovoltaic (PV) panels needs the knowledge of solar radiation at that particular area, so due to lack of funds, a method to approximate the ground level solar radiation for a given space position on earth, during a specific period, using a deterministic solar radiation computation and available experimental data has been employed. Computer modelling studies of Dye-Synthesized Solar Cells (DSSCS) is also discussed.

2. Materials and Methods

2.1 Study Area

The overview of the solar energy landscape is done in Limpopo, which is one of the nine provinces of SA. This province is situated in the far north of Gauteng the capital city of SA. SA has abundant of sunshine throughout the year. It has one of the highest insolation rates in the world, between 4.5 kWh/m^2 and 6.5 kWh/m^2, and receives about 2 500 hours of sunshine a year (over 300 days of sunshine per year in some provinces like Limpopo) [Munzhedzi, *et al.*, 2009].

Limpopo is the gateway of many African countries like Zimbabwe, Mozambique, Botswana and Namibia. The population in Limpopo is now 5 700 090. Most of its areas are rural comprised of non-employed people. So, they experience difficulties in paying electricity from ESKOM. It is the fifth's largest of the nine provinces covering 125 755 km^2. The northern and eastern areas are subtropical with mild and mostly frost free winters [www.brandsouthafrica.com/tourism-south-africa/

geography/limpopo].

Figure 12.1: Provincial Map of South Africa.

2.2.1 Available Solar Parks in limpopo

The department of Energy (DoE) has committed to increase renewable generation (wind, solar, hydro-power, geothermal, *etc.*) by 13 225 MW in 2025. The government is trying its level best to change SA to be green. Currently this clean energy is contributing 4.5 per cent of SA's 44 000 MW of generating capacity [State of Renewable Energy in South Africa 2015]. The use of solar energy in Limpopo is in the form of photovoltaic panels tied to the grid as well as off-grid. Three operational solar plants are Witkop Solar Park, Soutpan Solar Park and Tom Burke solar plant. These are all grid-tied.

2.2.2 Witkop Solar Park

The development of this plant falls under window 1 of the government. It is situated west of Polokwane city. The PV panels are mounted to the solar trackers. This solar plant was commissioned in September 2014 and it has the capacity of 30 MW.

2.2.3 Soutpan Solar Power Plant

The plant is just 2 km from Vivo. Commercial operation commenced in July 2014 and generates 132 kV with the capacity of 28 MW through 108 000 PV panels that are mounted to solar tracker too. It is connected to the ESKOM existing substation. Witkop and Soutpan solar power plants serve about 50 000 rural households [5].

Figure 12.2: Witkop Solar Power Plant.

Figure 12.3: Soutpan Solar Power Plant.

2.2.4 Tom Burke Solar Plant

This plant is situated at Lephalale and it has covered about 148 hectares and provides 66 MW capacity. Annual production is 122 GWh which serves about 38 000 households. Utilization of renewable energy brings a solutions of the reduction of carbon footprint. About 111 000 tonnes of carbon dioxide is reduced per year [www.observer.co.za/limpopos-third-solar-plant-in-operation].

Figure 12.4: Tom Burke Solar Park.

2.3 Small Scale - PV Panels in Limpopo

Households, clinics, schools, *etc.* are privately using solar energy to reduce the burden of finance. This energy is mostly used for lighting and some electricity appliances of few watts like televisions and radios. Such households prefer the off-grid tied such as the picture attached below. There are a number of households that are connected off-grid in the deep rural areas.

Figure 12.5: Standalone PV.

The 20 X 255 W panels equivalent to 5100 W solar plant is installed at Vuwani Science Resource centre under the University of Venda with capacity of 5100 W in

Figure 12.6: Vuwani Resource Science Centre Solar Power.

2016 and supplies solar energy to a building of four laboratory rooms and three offices. It uses a 5 kVA grid tied inverter. The supply of this energy is used during the day though the excess is fed to the grid without rebates and in the evening Eskom takes over. There is a reduction of the monthly Eskom charges. The graduate students who are working at the centre as interns are gaining skills, such analysis of the data retrieved from the solar frontier.

2.4. Theoretical Determination of Solar Energy in some Regions in Limpopo

An extensive radiation measurement of high quality of solar energy is very crucial. Due to lack of funds, installation of radiometers in each region becomes a problem. So, for the determination of solar radiation, the empirical solar models that use the meteorological data is an alternative way. Though most of these empirical models were not developed in South Africa, the researchers use the available data, such as temperature, sunshine, humidity, *etc.* to develop or modify the models and then employ them to determine either the global, direct or diffuse solar radiation. Preferably, the calculated solar energy should be in kilojoule (kJ) or mega Joule (MJ). In this study, Mutale, Marblehall and Ammondale areas were considered whereby Angstrom linear regression constants and global solar radiation using Hargreaves and Samani model were determined.

Angstrom linear model [Angström, A., 1924]

$$\frac{H}{H_0} = \alpha + \beta \overset{x}{*} \left(\frac{S_a}{S_p}\right) \tag{1}$$

Hargreaves and Samani model [Allen, R., 1995,1998]

$$\frac{H}{H_0} = k_r \dot{n}\left(\sqrt{T_{max} - T_{min}}\right) \tag{2}$$

where, H is the global solar radiation in MJ/m^2, H_0 is the extraterrestrial solar radiation above the earth surface, α and β are linear regression coefficients, k_r the empirical constant estimated by Samani where its value is 0.16 for interior region or 0.19 in the coastal region, T_{max} – maximum temperature and T_{min} – minimum temperature measured at that particular site. The relation [Duffie, J.A. and Beckman, W.A.,1991],

$$H_o = \frac{24}{\pi}(1367\ W/m^2)\left(1 + 0.33cos\left(\frac{360n}{365}\right)\right)$$

$$\left(\frac{\pi}{180}\omega_s \sin\varphi \sin\delta + \cos\varphi \cos\delta \sin\omega_s\right) \tag{3}$$

determines the extraterrestrial solar radiation in MJ/m^2 with

$$\omega_s = \cos^{-1}(-\tan\ \phi\ )(\tan\delta) \tag{4}$$

$$\delta = 23.45 \sin\left(360\left(\frac{284+n}{365}\right)\right) \tag{5}$$

where δ – the declination in degrees.

Though some of the weather stations do measure the global solar radiation, in this study, we used equation (2) to estimate H so as to compare with the in-situ data. The comparison of the in-situ and estimated H were compared to verify the suitability of the model in the South African weather. The models use the angstrom regression coefficients and the measured sunshine hours as illustrated on equation 1. The linear regression coefficients constants vary depending on the weather conditions at the study site, S_a - the daily actual number of sunshine hours and S_p - the possible sunshine hours or day-length. This can be calculated from the following relation;

$$S_p = \frac{2\omega_s}{15} \tag{6}$$

2.5. Theoretical Modeling of a Novel Material for the Application in Solar Cell

Solar energy conversion is part of a long term strategy to ensure a stable and adequate supply of electrical power in the future. Photovoltaics are the only method of converting sunlight directly into electrical energy. The efficiency of a photovoltaic system is measured as the ratio of electrical power produced to the energy of the incident solar radiation. It strongly depends on the quality of the semiconducting materials used for the fabrication of solar cells. Presently, the photovoltaic market is dominated by silicon technologies, however, the challenge is to manufacture more cost effective solar cell materials and maintain a high efficiency.

Experimental work for the development of novel material for application in solar cells is expensive due to the fact that there are many reagents involved, and without a proper method the cost increase because of all the trial to be conducted to get the right material. Hence the use of Theoretical modeling of novel material for application in solar cells comes in handy, because this theoretical method is computer based models which makes the less expensive to do and first to process. The results from these theoretical and computer based models can help the experimental scientists with a proper synthetic root with less trials.

In our group we are modeling materials to be used in two of the most promising solar cells at low cost, which are the dye sensitized solar cells and the chalcopyrite-type thin films. For dye sensitized solar cells we are working with titanium dioxide (TiO_2) as a semiconductor. This TiO_2 semiconductor has a drawback of absorbing sunlight in the ultraviolet region which occupies a small (5 per cent) part of the electromagnetic spectrum, meanwhile the visible region occupies the larger (45 per cent) part of the electromagnetic spectrum, this is due to the large band gap of TiO_2 which is about 3.4 eV. In order to reduce this band gap and shift the absorption of light to the visible region we do same band gap engineering and also introducing same element in the TiO_2 framework which are known to have the ability to narrow the band gap and shift the absorption of light.

A typical chalcopyrite-type thin-film solar cell consists of a p-type, chalcopyrite-type absorber, and an n-type transparent window layer which together form a p-n junction. The window layer consists of a CdS buffer layer and a ZnO front contact. These layers with thicknesses of about 2 μm are all deposited on molybdenum-coated substrates as, *e.g.*, glass or flexible Ti foils. The Mo layer acts as the back contact of the solar cell. In addition to the layers mentioned above, a $MoSe_2$ layer is visible between Mo and $Cu(In,Ga)Se_2$, formed during the coevaporation of Cu, In, Ga, and Se.For research and development of the absorber growth, we analyse the effect of parameter variations (*e.g.*, deposition temperature or composition) on the microstructure of the absorber and to compare this information with the solar-cell performance.

All these calculations for both the dye sensitized solar cells materials and chalcopyrite-type thin-film solar cells materials are carried out using CASTEP (Cambridge Sequential Total Energy Package) code in Materials Studio of Accelrys Inc.

3. Results and Discussions

3.1 Generation of Solar Energy and Regression Models

South African government is doing its best in establishing alternative systems for the supply of electricity. The annual efficiency of energy is increasing. This is evidence through the new solar plants, wind or biomass plants that are developed.

Table 12.1 shows good comparison of the observed and estimated global solar radiation from the year 2007 to 2011 whereby equation (2) was employed. Acceptable minor difference $|H_{est} - H_{obs}| \leq 2.5 \text{ MJ/m}^2$ is observed. This concludes that,

Samani and Hagreaves model is suitable for estimating global solar radiation in South Africa.

Table 12.1: Ammondale - Observed and Estimated Global Solar Radiation, H

Month	Obs. 07	Est. 07	Obs. 08	Est. 08	Obs. 09	Est. 09	Obs. 10	Est. 10	Obs. 11	Est. 11
Jan	20.2	21.8	16.7	19.3	21.9	19.7	23.0	20.1	21.7	18.9
Feb	20.7	22.5	19.1	21.4	21.8	19.2	23.6	20.0	22.3	20.6
Mar	18.5	19.9	14.7	18.0	20.5	18.0	20.5	19.1	22.0	19.9
Apr	13.8	16.7	14.1	18.2	20.4	18.4	13.6	13.9	14.9	15.4
May	14.8	17.2	13.0	15.5	16.4	15.2	15.7	14.7	15.6	15.6
Jun	13.9	14.9	11.9	14.5	14.7	13.9	15.4	14.6	16.0	15.3
Jul	14.7	15.8	12.1	15.1	16.2	15.2	14.8	14.8	15.2	15.7
Aug	16.9	18.7	15.4	18.8	18.8	18.3	18.4	18.9	19.2	18.4
Sep	17.8	20.0	17.4	21.9	21.2	20.8	20.8	21.8	22.8	22.4
Oct	15.7	19.1	24.1	22.2	23.1	20.5	22.2	21.8	23.1	22.6
Nov	16.5	20.7	9.2	7.9	23.7	20.4	19.5	21.0	22.6	21.9
Dec	17.5	19.1	23.3	20.5	26.9	21.7	22.8	21.1	20.7	20.7

Though the computation of the regression coefficients were done for six years, 2007 – 2012 with the use of MATLAB interface, few graphs are shown as Figures 12.7 and 12.8 for Ammondale and Mhinga areas. These regression coefficients are very useful because for the sites where the solar measurements are not available, Angstrom model can be employed for the determination of global solar radiation. From the graphs, it can be deduced that in most of the days of the year, the clearness index lies between 0.8 and 1, this actually confirms that South Africa experience more sunshine days. That is approximately 80 per cent of the number of days, there is less clouds.

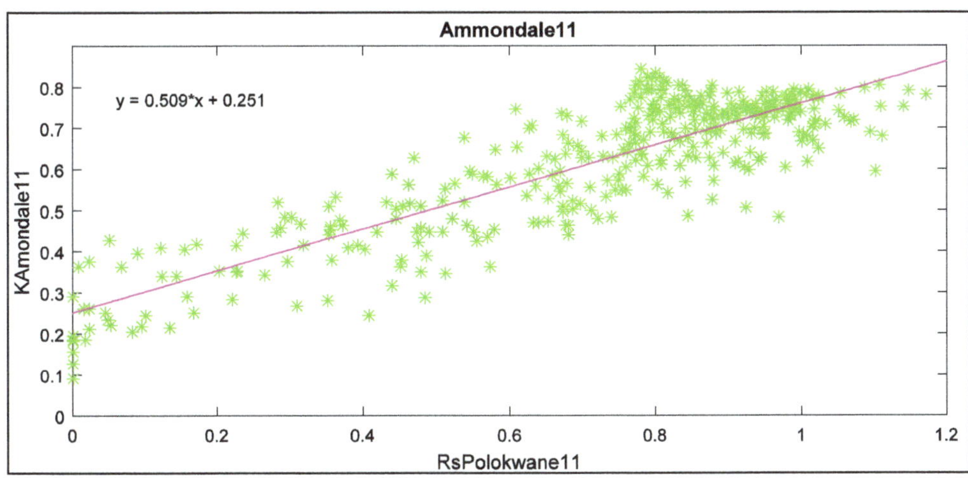

Figure 12.7: Ammondale - Relative SS versus Clearness Index.

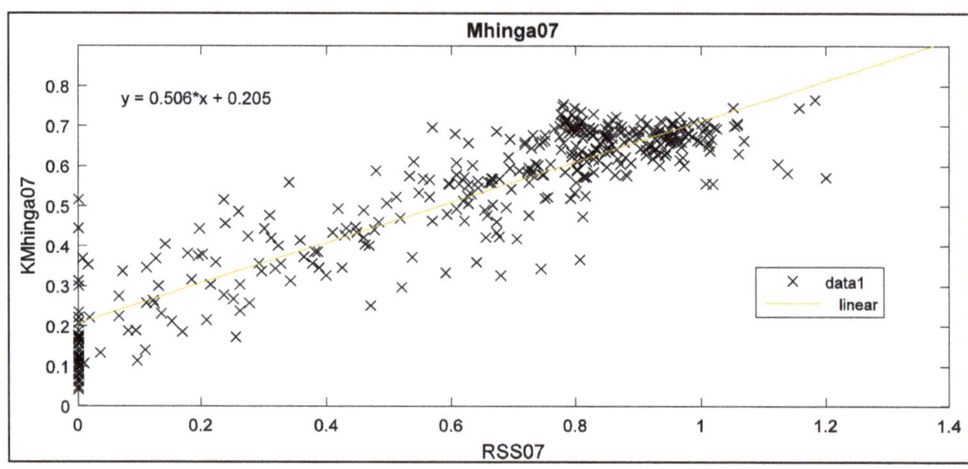

Figure 12.8: Mhinga - Relative SS versus Clearness Index.

Figures 12.9 and 12.10 depict the comparison of the estimated and the measured global solar radiation versus the days of the year in Marble Hall, Limpopo Province South Africa. It is evident from the graphical representations that the observed H (blue) and the calculated H (red) are in good comparison. The calculated H were computed through equation 2. One can summarize by saying that wherever there are the temperature measurements, it is advisable to use Hagreaves and Samani model to estimate the amount of global solar radiation.

3.2. TiO_2 Modelling Results

3.2.1. Surface Properties

The surface structures of brookite have been modeled by cleaving the optimized

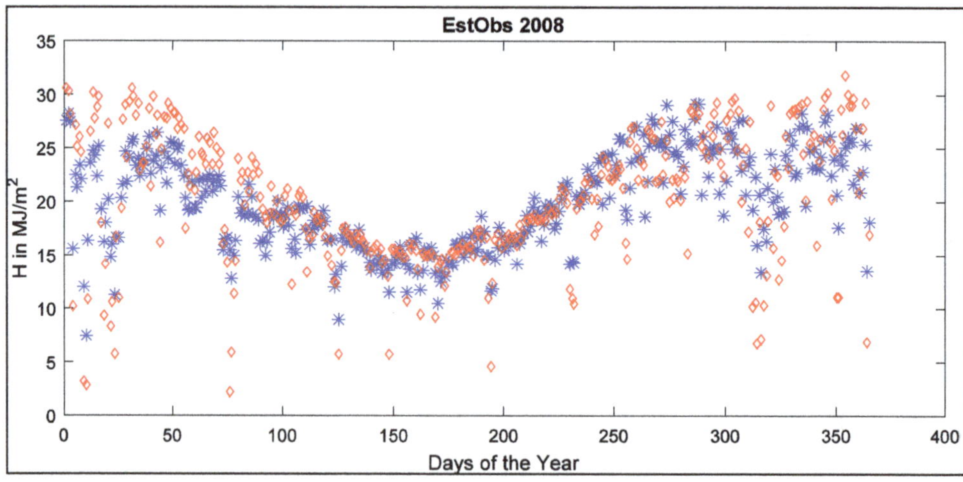

Figure 12.9: Marble Hall – Comparison of the Observed and Estimated H.

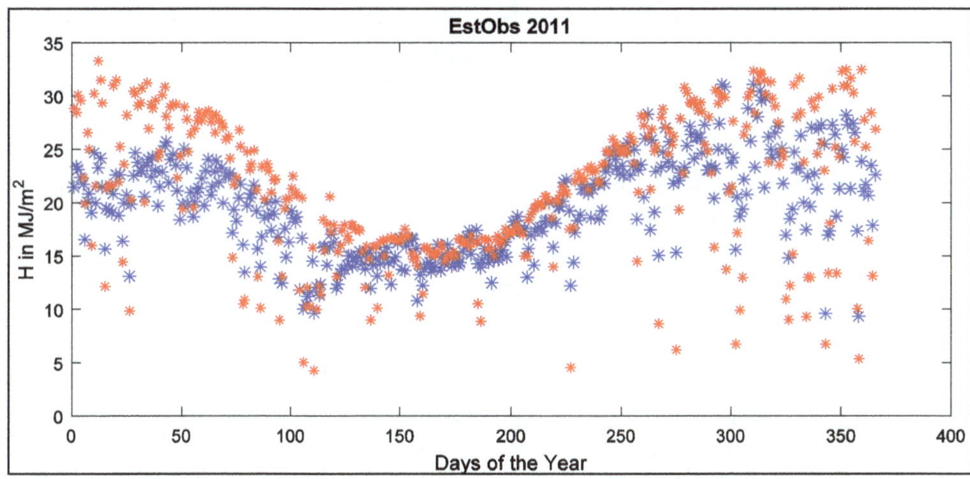

Figure 12.10: Marble Hall – Comparison of the Observed and Estimated H.

bulk structure and then build a slap for all surfaces. As these surfaces have a different number of atoms in each layer. The relaxation of these surfaces has been taken into account by optimizing all the layers of the slab.

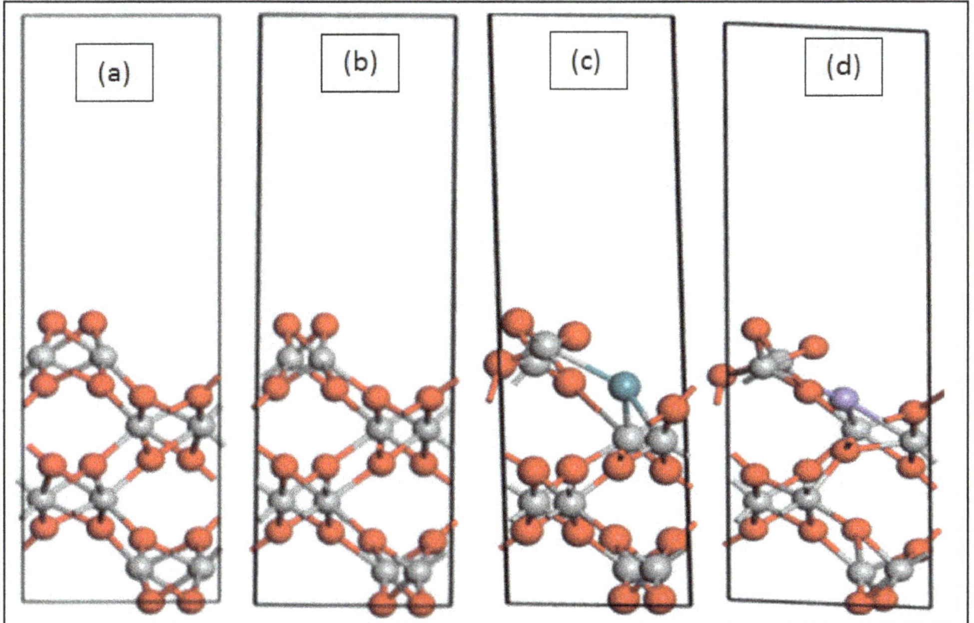

Figure 12.11: Atomic Structures of (a) (100) Un-doped TiO_2 Brookite Surface before Optimization, (b) (100) Nu-doped TiO_2 Brookite Surface after Optimization, (c) (100) Ru-doped TiO_2 Brookite Surface After Optimization, (d) (100) Fe-doped TiO_2 Brookite Surface After Optimization.

A schematic representation of the brookite surfaces is depicted in Figure 12.11. The truncation of the octahedral renders different coordination combinations for the outermost titanium cations. Depending on the slice of the brookite surface, the surface may be either oxygen- or titanium terminated. But in this work, we didn't consider the termination site. All the (100) brookite surface structures where optimized by relaxing surface atoms to eliminate surface atomic tension. The dopants were introduced by replacing one 3-fold oxygen element.

The surface atomic position is adjusted to reduce the total energy of the system. Figure 12.11 shows the structures of the un-doped and doped brookite surfaces before and after optimization, the atomic surface underwent relaxation, but not reconstruction. For all the surfaces the top and bottom layers of oxygen ions are still 2-fold coordinated (O_{2c}), with the nearest titanium ions being 5-fold coordinated (Ti_{5c}). Table 12.2 below illustrate the structural parameters of all four structures after optimization. It can be observed that the lattice the "a" and "b" parameters of doped structures doesn't have much difference with that of the doped structures, but the lattice parameter "c" for the un-doped structure is lower than that of Ru and Fe-doped structures "c" lattice parameters meanwhile the "c" of the dopants are close to each other.

Table 12.2: Optimized Structural Parameters for Surface TiO_2 Compared with Experimental and Previous Theoretical Results

	Un-doped TiO_2	*Ru- doped TiO_2*	*Fe-doped TiO_2*
a (E)	5.558	5.581	5.491
b (Å)	5.500	5.447	5.555
c (Å)	18.556	19.903	19.362

3.2.2 Electronic Properties

To conveniently investigate the electronic structures of transition metal-doped brookite TiO_2 surfaces, we set the same k-points mesh to sample the first Brillouin zone for pure and transition metal-doped models.

The calculated band gap of the (100) brookite TiO_2 surfaces were found to be underestimated compared with the experimental; this is due to the limitation of DFT. Which is the discontinuity in the exchange correlation potential which is not considered within the framework of DFT.? However, our discussions about energy gap will not be affected because only the relative energy changes are of concern. Even though the limitation of the DFT won't affect the discussion of the results, in this work we employed the scissor operation of 1.047 eV to compensate for the underestimation of the band gap. The Scissors Scheme align both the theoretical VBE and CBE with their experimental counterparts, performing a "scissors" operation to stretch out the theoretical gap states over the experimental gap. The value of the "scissors" operation used for these surfaces is the same as the one obtained in the bulk structure.

The calculated band gap of pure (100) brookite TiO_2 surface is 3.350 eV as shown in Figure 12.14. The conduction band minimum (CBM) is located at G, while the

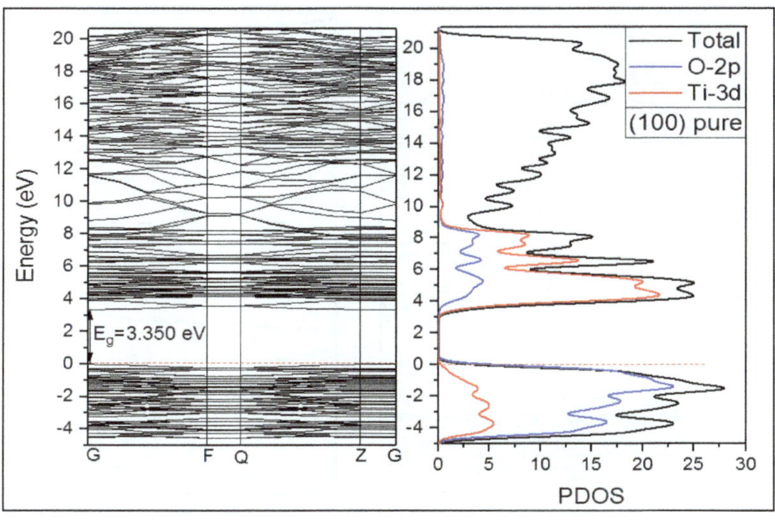

Figure 12.12: Band Structure and State Density of Un-doped Brookite TiO_2 (100) Surface.

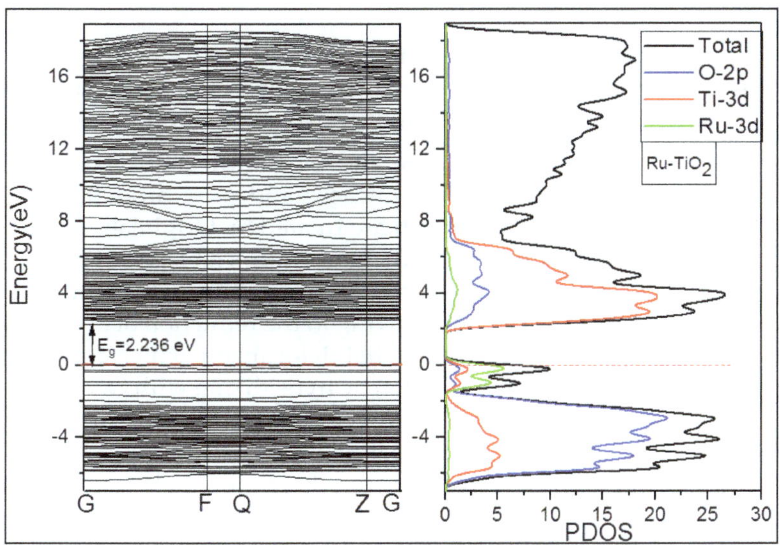

Figure 12.13: Band Structure and State Density of Ru-doped Brookite TiO_2 (100) Surface.

valence band maximum (VBM) is also located near G. So, the brookite TiO_2 (100) surface can be considered as a direct band gap semiconductor. The value of band gap is consistent with the reported results [1]. The valence band energy in the pure surface near the Fermi level is from 0 to 4.4 eV and the width is 4.4 eV, which is mainly attributed to the O-2p orbit and Ti-3d orbit hybridization. The CBE is from 2.5 to 20 eV and the width is 17.5 eV, mainly constituted by the Ti-3d track and a small amount of O-2p track. After doping of transitional metal the band gap narrowed,

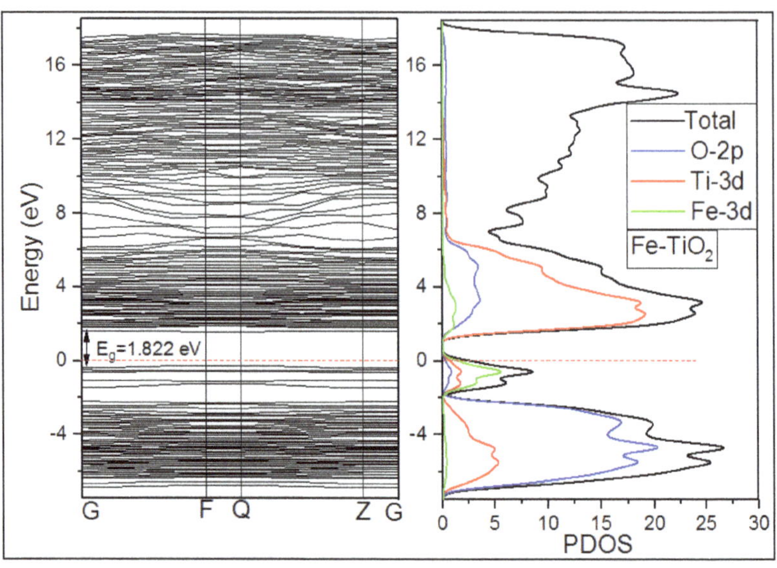

Figure 12.14: Band Structure and State Density of Fe-doped Brookite TiO_2 (100) Surface.

which is in support to the results reported by Wang *et al.*, who reported that most transition metal doping can narrow the band gap of TiO_2, lead to the improvement in the photo reactivity of TiO_2, and simultaneously maintain strong redox potential.

The total density of states (TDOS) and partial density of states (PDOS) of transition metal-doped brookite TiO_2 surfaces and pure brookite TiO_2 surfaces are shown in Figures 12.14–12.16, which are treated by Originlab. The band gap is defined as the separation between the VBM and CBM. The TDOS shape of transition metal-doped TiO_2 becomes broader than that of pure TiO_2, which indicates that the electronic nonlocality is more obvious, owing to the reduction of crystal symmetry. The transition metal 3d states are somewhat delocalized, which contributes to the formation of impurity energy levels (IELs) by hybridizing with O-2p states or Ti-3d states. Such hybrid effect may form energy levels in the band gap or hybrid with CBM/VBM, providing trapping potential well for electrons and holes. It gives a contribution to separation of photo-generated electron–hole pairs, as well as in favour of the migration of photo excited carriers and the process of photocatalysis. The band gaps of TiO_2 doped with Ru and Fe are determined as 2.236 eV, and 1.822 eV, respectively. Similar results were reported by Hanaor *et al.*, 2012 who investigated the Ab Initio study of phase stability in doped TiO_2 and reported that the Ab-initio density functional theory (DFT) calculations of the relative stability of anatase and rutile polymorphs of TiO_2 which were carried out using all-electron atomic orbitals methods with local density approximation (LDA). They found that Al and Fe dopants were found to act as shallow acceptors with charge compensation achieved through the formation of mobile carriers rather than the formation of anion vacancies [H. Kominami *et al.*, 2000].

3.2.3 Optical Properties

The polycrystalline model of TiO_2 was used to calculate the absorption spectra of the different systems. Because GGA method has the disadvantage of band gap underestimation, we used a "scissor operation" of 1.047 eV. The optical calculations were based on the ground state of the electrons. The calculated absorption spectra for pure, Ru doped, and Fe doped TiO_2 are shown in Figure 12.15.

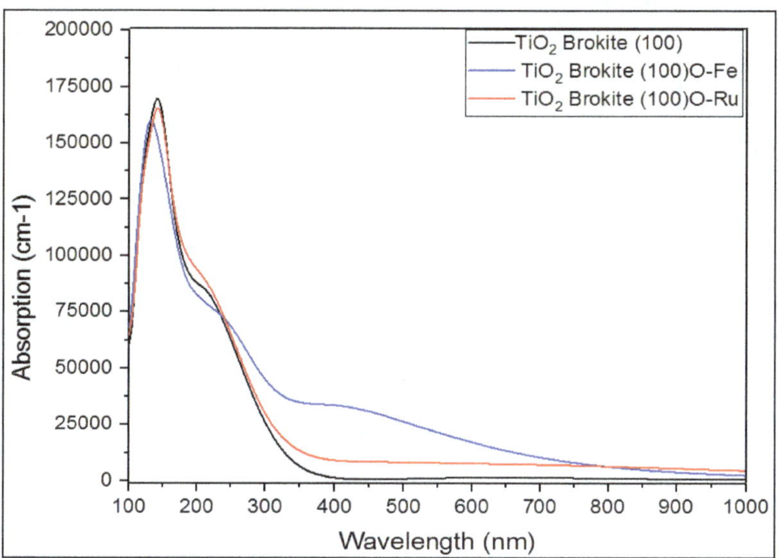

Figure 12.15: Optical Absorption Curves of the Un-doped and Doped Brookite TiO_2 (100) Surface.

From Figure 12.15 it can be seen that the absorption edge of all the doped TiO_2 shifts to visible light region with respect to pure TiO_2, and Fe doped TiO_2 has a greatest red shift extent. This can be observed in Figure 12.17 with Fe-doped surface having a high peak absorbing from 400nm towards the infrared (IR) region. As all the doped TiO_2 have a lower band gap than pure TiO_2, the red shift should be caused by impurity energy levels. The empty impurity energy levels below conduction band form a new conduction band minimum (CBM), and the occupied impurity energy levels above valence band act as new valence band maximum (VBM), so the absorption edge shifts to visible light region. The optical absorption shows exactly were in the electromagnetic spectrum does element absorb light at, or were in the electromagnetic spectrum there is photocatalytic activity of a given semiconductor.

The real and virtual parts of the dielectric function versus photon energy change curve are shown in Figures 12.16 and 12.17. For the undoped TiO_2 and Ru-doped there are four peak positions in the imaginary part. The first high and sharp peak for both the un-dope and Ru-doped appears at ~4.8 eV this is because the valence band O2p states jump to the conduction band Ti 3d states but for the Ru-doped there are some contribution from Ru 3d state, according to analyses of the energy band diagram and density of states. The second, third and fourth peaks are at ~7

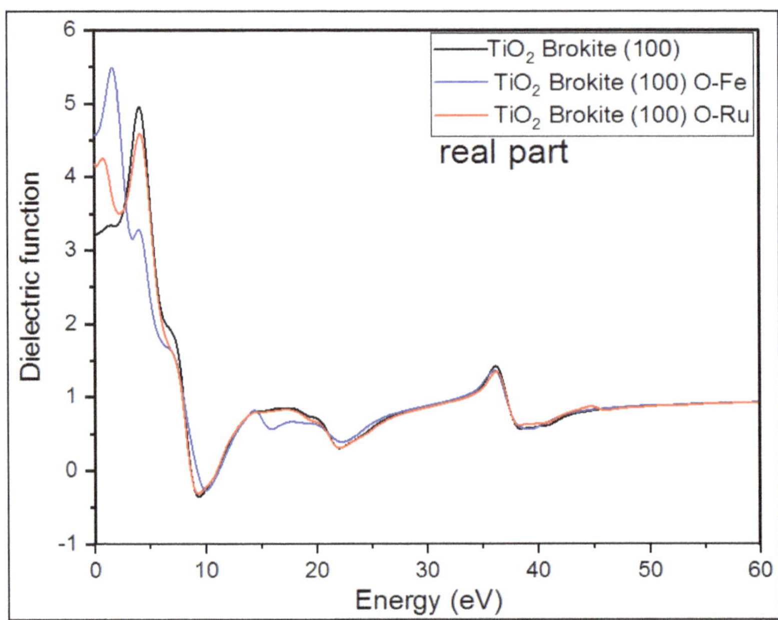

Figure 12.16: Dielectric Function of the Un-doped and Doped Brookite TiO_2 (100) Surface.

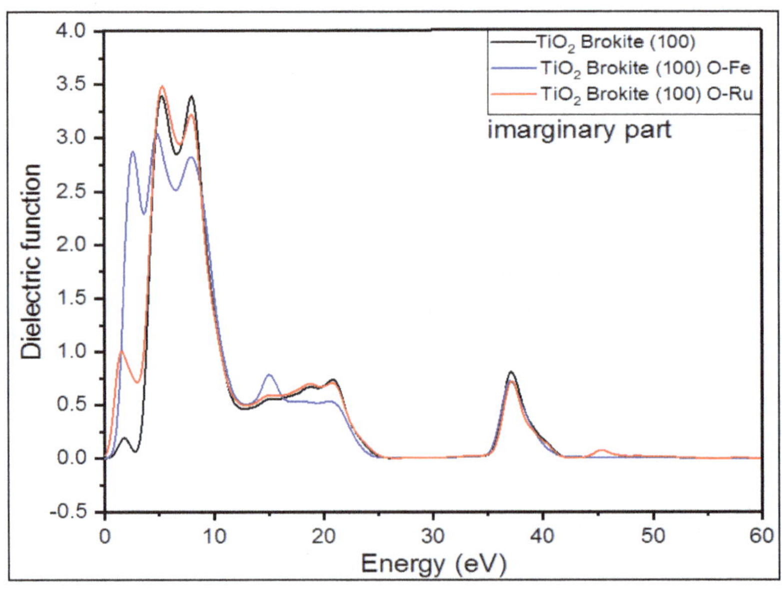

Figure 12.17: Dielectric Function of the Un-doped and Doped Brookite TiO_2 (100) Surface.

eV, ~22 eV and ~38 eV mainly owing to the electron transition from the O 2s state to the Ti 3d state. There were no results in literature to compare with. For Fe-doped

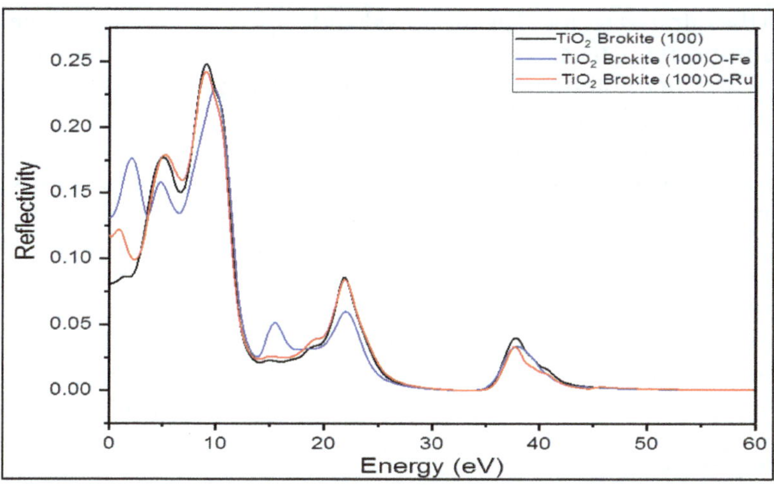

Figure 12.18: Reflection Spectrum of the Un-doped and Doped Brookite TiO_2 (100) Surface.

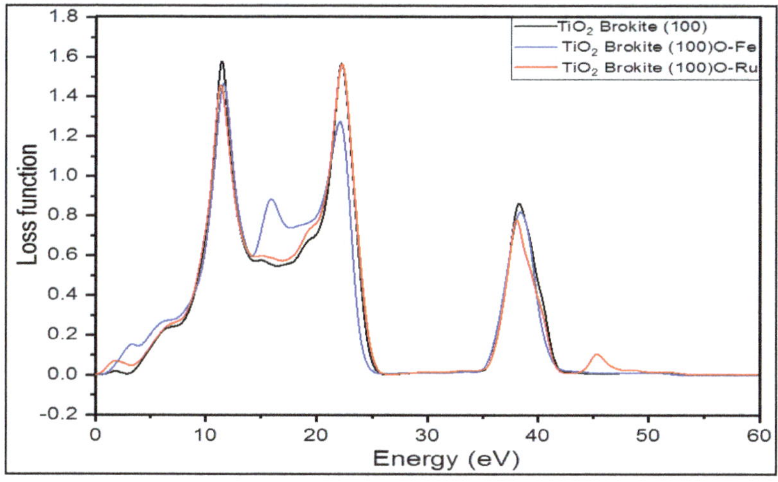

Figure 12.19: Energy Loss Function of the Un-doped and Doped Brookite TiO_2 (100) Surface.

both real and imaginary parts of the dielectric function has some number of peaks. The peak of the real part which is due to the valence band O 2p states jump to the conduction band Ti3d states and some contribution from Fe 3d state, according to analyses of the energy band diagram and density of states. This peak at low energy than that of un-doped and Ru-doped. The second peak of Fe-doped in both real and imaginary parts is showing some splitting at the top forming another broad peak.

The reflectivity is the ratio of a wave reflected from a surface of energy possessed by the wave striking by the surface. The reflectivity spectra's of un-doped and doped surfaces are displayed in Figure 12.19 as a function of energy. The reflectivity is much higher in the infrared region. The un-doped and doped surfaces are showing similar reflectivity spectra patens, where the first peak is at low energy reigning between 0_10 eV, with some splitting at the top. The other two peaks are at ~20 eV and ~38 eV, with decreasing intensity. For Fe-doped the reflectivity spectra is showing four peak, where three are in the same energy region as those of un-doped and Ru-doped, but with intensity or low photo energy in some regions. The other peak which only appear in Fe-doped is at 15 eV with a well define shape and low intensity. The large reflectivity for E < 1ev indicates the characteristics of high conductance in the low energy region. The energy loss spectrum describes the energy loss of electrons when passing through a uniform dielectric, and the energy loss peak describes the plasma resonance frequencies. The main energy loss for all the surfaces is at about ~22 eV and ~38 eV. Figures 12.16 and 12.17 reveal that the energy loss peaks for all surfaces correspond to the sharp decline in the reflectivity spectrum.

4. Conclusions

The South African renewable energy strategy is carried out in the provincial level as well as in the country as a whole. A number of solar plants are commissioned as planned. It is evident that by 2030, the country will be able to reach the target of producing 52 GW through renewables.

The regression coefficients calculated modifies the Angstrom linear regression model as

$$\frac{H}{H_0} = 0.231 + 0.501\frac{S_a}{S_p}$$

If $\frac{S_a}{S_p} = 1$, $\frac{H}{H_0} = 0.732$; which shows that, Limpopo does receive more than 70 per cent of solar irrespective of season.

The un-doped and Transition metals (Ru and Fe)-doped TiO_2 has been studied using first-principles density functional theory. We used the generalized gradient approximation (GGA) in the scheme of Perdew-Burke-Ernzerhof (PBE) to describe the exchange-correlation functional. All calculations were carried out using CASTEP (Cambridge Sequential Total Energy Package) code in Materials Studio of Accelrys Inc. Based upon calculated results, the analysis of electronic structures suggest that, metal dopants shift the VBs to higher energy all surface. The results also show that owing to the formation of the impurity energy levels, which are mainly hybridized by 3d states of impurities with O 2p states or Ti 3d states, the response region in spectra could be extended to the visible light region. The components of VBs, CBs and dopant states support that the metal dopants are active in inter-band transitions induced by lower energy excitations, which is important for the application of solar energy.

The compensating optical absorbance of Ru and Fe dopants in different wavelength region present that the both dopants may possess prominent optical absorption properties. These doping systems can be easily obtained and with good stability. Theoretical research on transition metal-doped TiO_2 is of great importance to develop the photocatalytic applications. First-principles calculation of doped TiO_2 is still an ongoing subject, and a few challenging problems require further investigation in an urgent demand. One is the influence of the transition metal doping on the phase transition of TiO_2 from anatase to rutile. A theoretical understanding on its mechanism will be useful to optimize the performance of TiO_2 in photocatalytic and other applications. Another one is the question about using the virtual crystal approximation method to calculate the doping system for very low concentration, which can cut down the calculation time. With the solution of these problems, one could provide more accurate theoretical models to simulate the practical doping approaches which could lead to important implications in the optimization of the performance of transition metal-doped TiO_2 photocatalysts.

5. Acknowledgements

I would like to express my gratitude to the organizers of NAM S&T CENTRE for the financial support they have given me to be able to attend this valuable workshop. I am obliged to the University of Venda, who accepted my request to be away for this period so as to be able to participate in this workshop. It would really be unfair if I don't acknowledge the Renewable energy group in our department who gave invaluable support for the completion of this work. I sincerely thank the department of Physics staff who encouraged me during the preparation of the work.

Above all, I thank God for the strength He gave me through all obstacles I came across.

6. Abbreviations

SA: South Africa

PV: Photovoltaic

MW: Megawatt

MJ: Megajoule

W: Watt

BAU: Business As Usual

DoE: Department of Energy

DSSCS: Dye-synthesized solar cells

H: Global solar radiation

w_s: Sunset hour angle

S_a: Daily actual number of sunshine hours

S_p: Possible sunshine hours

CB: Conduction Band

VB: Valence Band

GGA: Generalized gradient approximation

CASTEP: Cambridge Sequential Total Energy Package

DFT: Density Function heory

REFERENCES

1. Angström, A., 1924. Solar and terrestrial radiation. Q.J.R. Meteorol. Soc. 50, 121–125.
2. Allen, R., 1995. Evaluation of procedures of estimating mean monthly solar radiation from air temperature. FAO, Rome.
3. Allen, R., 1997. Self-calibrating method for estimating solar radiation from air temperature. J. Hydrol. Eng. 2, pp. 56–67.
4. Community Survey 2016, Statistical release P0301/Statistics South Africa (http://www.statssa.gov.za) accessed 12 March 2017.
5. Duffie, J.A. and Beckman, W.A. Solar engineering of thermal processing, 1991, 2nd ed. Madison: John Wiley and Sons, Inc.
6. Hanaor A.D, *et al.*, 2012. Ab Initio Study of Phase Stability in Doped TiO2. Comp. Mech., 2, pp. 185-194.
7. Kominami H,et al 2000. Synthesis of brookite-type titanium oxide nano-crystals in organic media. J. Mater. Chem., 10, pp. 1151.
8. Munzhedzi, R., *et al.*, 2009. Redrawing the solar map of South Africa for photovoltaic applications, Renewable Energy, 34, pp. 165-169.
9. State of Renewable Energy in South Africa 2015 published by the Department of Energy.
10. http://www.brandsouthafrica.com/tourism-south-africa/geography/limpopo (accessed 13 March 2017).
11. http://www.observer.co.za/limpopos-third-solar-plant-in-operation (accessed 13 March 2017).

Chapter 13

Fabrication of CdS/CdTe Thin Film Solar Cells via the Technique of Electrodeposition

***H.Y.R. Atapattu*[1], *D.S.M. De Silva*[1]*, *A.A. Ojo*[2] *and I.M. Dharmadasa*[2]**

[1]Department of Chemistry, University of Kelaniya, Kelaniya, Sri Lanka
E-mail: sujeewa@kln.ac.lk
[2]Materials and Engineering Research Institute, Sheffield Hallam University, Sheffield S1 1WB, UK

This study focused on fabrication of CdS/CdTe solar cells using the technique of electrodeposition as it is simple, low cost and scalable method. Initially, CdS and CdTe materials were individually deposited on fluorine doped tin oxide (FTO) glass substrates and optimum growth conditions were obtained by analyzing their structural, compositional, electrical, optical and morphological properties using the techniques of X-ray diffraction, Energy Dispersive X-ray spectroscopy, photo-electrochemical cell study, optical absorption spectroscopy and scanning electron microscopy respectively. Thereafter, final device structure ofglass/FTO/CdS/CdTe/Au was fabricated using the optimum growth conditions obtained for the two materials, CdS and CdTe. Finally the current density-voltage characteristics of the devices were obtained to assess devices. The best device structure exhibited short circuit current density (J_{sc}) of 24.4 mA cm^{-2}, open circuit voltage (V_{oc}) of 681.9 mV, Fill Factor (FF) of 0.32 and conversion efficiency of 5.4 per cent.

Keywords: *Cadmium sulfide, Cadmium telluride, Thin films, Solar energy materials, Electrodeposition, Solar cells.*

---

* Postal Address: Dr. D. S. M. De Silva, Department of Chemistry, University of Kelaniya, Kelaniya 11600, Sri Lanka; Email: sujeewa@kln.ac.lk; Phone: +94(0) 71 710 7497

1. Introduction

Recently, the global demand for energy has been tremendously grown up and already at a catastrophic level. Most of energy sources which are currently in use including fossil fuel are getting depleted as they are non-renewable energy sources(Kumar and Rao, 2014; Johnson, 2000; Atapattu *et al.*, 2016; Dharmadasa, 2013). In this regard, photovoltaic (PV) industry is a promising field to fulfill the current energy demand of the world at a comparably low cost (Kosyachenko, 2010). As a consequence, solar cell technology is firmly stepping forward while producing clean green energy to the future.

However, recent research exertions over the world intensely pay attention on further improving performances of solar cells and reducing the cost of production through further refining methodologies (Kumar and Rao, 2014; Kosyachenko, 2010; Romeo *et al.*, 2000). In this regard, CdS/CdTe based solar cells are being comprehensively explored due to their near ideal band gap for p-n junctions and high efficiencies (Kumar and Rao, 2014). On the other hand, the technique of electrodeposition (ED) is being investigated to produce CdS/CdTe device structures owing to its simplicity, low cost and scalability (Dharmadasa *et al.*, 2014). In this study an attempt is made to investigate the optimum growth conditions to deposit CdS and CdTe semiconductor thin films in order to fabricate low cost CdS/CdTe solar cells using the technique of electrodeposition.

2. Materials and Methods

Since the deposition technique of the materials of CdS and CdTe is electrodeposition throughout, the Metrohm Autolab PGSTAT128N potentiostat was used as the static potential source. To carry out all electrodepositions a three electrode electrolytic cell consisted of high purity graphite rod as the counter electrode and saturated calomel electrode as the reference electrode was used (see Figure 13.1). The constituents of the CdS aqueous electrolytic bath prepared were 0.10 mol/L $CdCl_2$ (Sigma Aldrich analytical grade $CdCl_2.H_2O$) and 0.01 mol/L $Na_2S_2O_3$ (Sigma Aldrich analytical grade $Na_2S_2O_3$). Correspondingly, CdTe aqueous electrolytic bath consisted of 1.00 mol/L $CdSO_4$ (Sisco research laboratory ACS grade $8CdSO_4.3H_2O$) and 150 mmol/L TeO_2 (sigma Aldrich 99.995 per cent TeO_2). Initially, CdS and CdTe layers were independently grown on fluorine doped tin oxide (FTO) coated glass substrates to identify cathodic deposition potential, pH of the electrolyte, deposition temperature, deposition time and stirring rate of the electrolyte. Subsequently, post deposition heat treatment was carried out to enhance the material quality of CdS and CdTe further.

Thereafter, the best conditions to grow PV active CdS and CdTe thin films were obtained by analyzing their structural, compositional, electrical, optical and morphological properties using the techniques of X-ray diffraction, Energy Dispersive X-ray spectroscopy, photo-electrochemical cell study, optical absorption spectroscopy and scanning electron microscopy respectively. Consequently, CdS and CdTe layers were electrodeposited in a stack on glass/FTO using the best growth conditions obtained for both materials and improved the PV quality of glass/FTO/CdS/CdTe device structure further by considering their electrical,

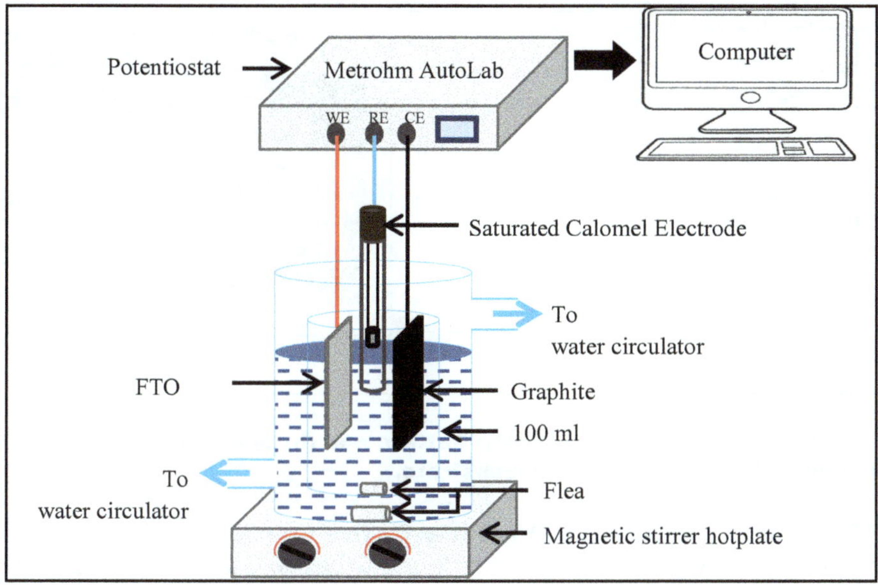

Figure 13.1: Schematic Diagram of the Electrodeposition Setup.

optical and morphological properties via the techniques that have been mentioned previously. Finally, the best glass/FTO/CdS/CdTe structures were conveyed to sputter back contact of Au (100 nm) and performance of glass/FTO/CdS/CdTe/Au devices were evaluated.

3. Results and Discussion

According to the obtainedstructural, electrical, optical and morphological properties of the material of CdS, the best conditions of the cathodic deposition potential, pH of the electrolyte, deposition temperature, deposition time and stirring rate of the electrolytic bath were identified to be in the ranges of 650-680 mV, 1.60-1.80, 55-65 C, 20-40 min and 60-125 rpm respectively. The best annealing temperature and time for CdS were found to be 400 C and 15 min respectively.

A typical X-ray diffraction pattern of a CdS layer electrodeposited on glass/FTO surface shown in Figure 13.2 (a) confirms the polycrystalline nature of the CdS sample as reported earlier (Dharmadasa, *et al.*, 2014; Mammadov *et al.*, 2012). According to the crystallographic pattern obtained, the CdS layer is mostly hexagonal (Greenockite) with a sharp diffractionpeakat 2θ-26.21° (002).

Figure 13.2 (b) illustrates the surface morphology of CdS thin film electrodeposited using the growth conditions optimized in this study. It shows the presence of uniformly scattered grains of CdS in the range of 100-250 nm. Moreover, the layer thickness of CdS is approximately in the range of 80-120 nm.

For growth of CdTe material, the best cathodic deposition potential, pH of the electrolyte and stirring rate of the electrolytic bath were identified to be in the ranges

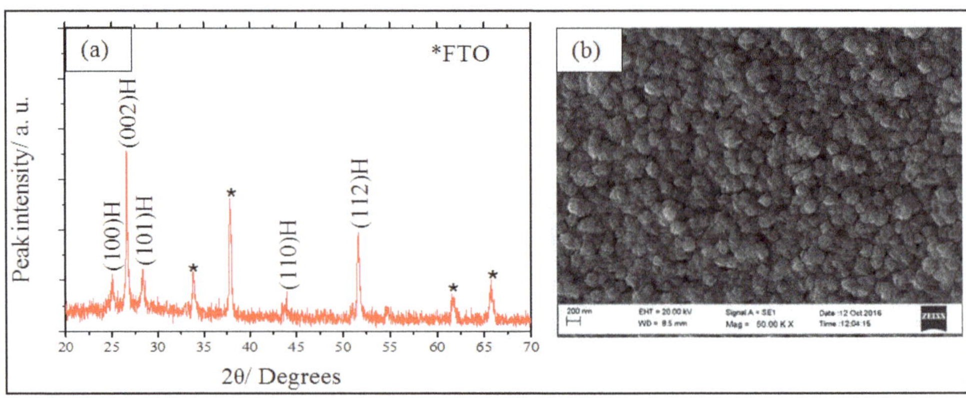

Figure 13.2: Typical X-ray Diffraction Pattern and (b) SEM Image of a CdS Layer Electrodeposited on Glass/FTO Substrate.

of 620-660 mV, 2.10-2.30 and 60-85 rpm respectively while annealing temperature in the presence of $CdCl_2$ and time were found to be 390 C and 15 min respectively.

A typical X-ray diffraction pattern of an as-deposited CdTe layer electrodeposited on glass/FTO substrate shown in Figure 13.3 (a) exhibits a cubic crystalline structure (Dharmadasa *et al.*, 2014). Further, the highest peak intensity can be observed along the crystallographic direction of (111) at 2θ of ~24 .

Surface morphology of an as-deposited CdTe layer shown in Figure 13.3 (b) represents the agglomerations of nano sized CdTe grains having average size of ~80 nm.

Finally, the glass/FTO/CdS/CdTe array was fabricated under the above optimized conditions and Au was sputter coated to form the back metal contact. The current density - voltage (J-V) characteristics of the best glass/FTO/CdS/CdTe/ Au device structure was measured and displayed in Figure 13.4.

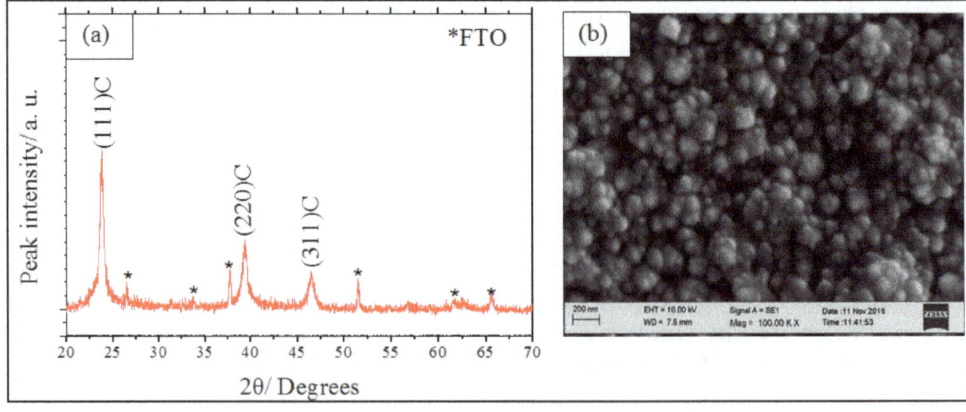

Figure 13.3: Typical (a) X-ray Diffraction Pattern and (b) SEM Image of an as-Deposited CdTe Layer Electrodeposited on Glass/FTO Substrate.

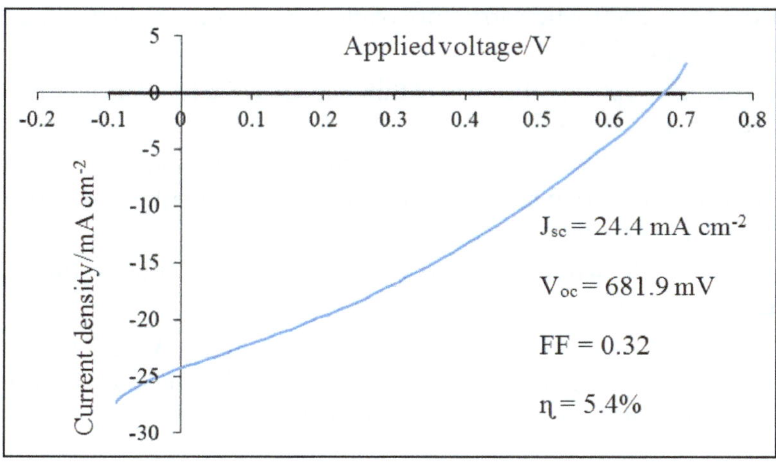

Figure 16.4: J-V Curve of the Glass/FTO/CdS/CdTe/Au Device Structure.

After the back contact (Au) fabrication, the best device exhibited the current density (J_{sc}) of 24.4 mA cm^{-2}, open circuit voltage (V_{oc}) of 681.9 mV and 0.32 of fill factor while producing 5.4 per cent solar energy conversion efficiency.

4. Conclusions

In this study, the technique of electrodeposition is successfully used to grow semiconductor materials of CdS and CdTe. Best growth conditions obtained were able to produce a solar energy conversion efficiency of 5.4 per cent with a considerable short circuit current density of 24.4 mA cm^{-2}. Further investigations on the interfaces of CdS/CdTe and CdTe/back contacts are needed for further improvement of open circuit voltage, fill factor and ultimately the conversion efficiency of solar cells.

5. Acknowledgements

Authors would like to acknowledge the University Grants Commission (UGC), Sri Lanka for their financial support under UGC Innovative Grants. Also, the contributions from O. I. Olusola and Muhammed Madugu at Sheffield Hallam University, UK are gratefully appreciated.

REFERENCES

1. Atapattu, H.Y.R., De Silva, D.S.M., Pathiratne, K.A.S. and Dharmadasa, I.M., 2016. Effect of stirring rate of electrolyte on properties of electrodeposited CdS layers. *J Mater Sci: Mater Electron*, 27, pp.5415–21.
2. Dharmadasa, I.M., 2013. *Advances in thin film solar cells*. Pan Stanford Publishing Pte. Ltd.
3. Dharmadasa, I.M. *et al.*, 2014. Fabrication of CdS/CdTe- based thin film solar cells using an electrochemical technique. *Coatings*, 4, pp.380-415.

4. Johnson, D.R., 2000. Microstructure of electrodeposited CdS/CdTe cells. *Thin Solid Films*, pp.321-26.

5. Kosyachenko, L., 2010. Efficiency of thin film CdS/CdTe solar cells. In Rugescu, R.D. *Solar Energy*. In Tech. Ch. 6. pp.105-30.

6. Kumar, S.G. and Rao, K.S.R.K., 2014. Physics and chemistry of CdTe/CdS thin film heterojunction photovoltaic devices: fundamental and critical aspets. *Energy and Environmental Science*, 7, pp.45-102.

7. Mammadov, M., Aliyev, A. and Elrouby, M., 2012. Electrodeposition of Cadmium sulfied. *Int. J. Thin Film Sci. Tec*, pp.43-53.

8. Romeo, N., Bosio, A., Tedeschi, R. and Canevari, V., 2000. Growth of polycrystalline CdS and CdTe thin layers for high efficiency thin film solar cells. *Materials chemistry and physics*, 66, pp.201-06.

Chapter 14

Advanced Materials in the Structure of Solar Cells

Khalil Azimeh

Department of Design and Production Engineering,
Damascus University, Syria
E-mail: kh_azim2000@yahoo.com

In industrialized countries, much attention is paid to the development of systems based on renewable energy sources, *i.e.* direct conversion of solar energy into electrical energy. For the effective operation of solar cells, the impedance included in series with the solar cell is necessary, the structure of the thin film must be uniform throughout the active region of the solar cell.

We considered the production of structures based on: 1-monocrystalline silicon; 2-amorphous silicon; 3-gallium arsenide; 4-polycrystalline semiconductors. We discussed the choice of materials for the substrate, as well as a more rational approach for the production of films. The creation of solar cells based on AlGaAs-GaAs heterostructures opened a new page in solar energy, as they were comparable in terms of weight and strength characteristics to silicon ones, and they outperformed them in terms of efficiency and radiation resistance.

Keywords: *Materials, Silicon, Thin-films, Polycrystalline, Polymer-fullerene, Anodium copper oxide, Solar cells.*

1. Introduction

The development of effective means of converting solar energy into electrical energy is an actual problem of today. Among the main advantages of solar energy can be identified the following: unlimited supplies of solar energy; Absence of harmful emissions into the environment; Decentralized production of energy, which excludes the creation of power lines; Absence of moving parts. The main shortcomings of solar energy should be attributed to the high cost of electricity received, as well as the local nature of the tasks to be solved.

In industrialized countries, much attention is paid to the development of systems based on renewable energy sources, including solar energy. One of the most attractive and promising renewable sources of energy has always been considered photovoltaic, *i.e.* direct conversion of solar energy into electrical energy. And also, according to estimates by 2050, solar energy can provide 20-25 per cent of world energy production, and by the end of the XXI century, solar energy should become the dominant energy source with a share reaching 60 per cent [1].

There are a number of conditions that must be available in order to obtain the effective operation of solar cells:

- ✰ The optical absorption coefficient (a) of the active layer of the semiconductor should be large enough to ensure absorption of an essential part of the energy of sunlight within the thickness of the layer;
- ✰ Electrons and holes generated by illumination should be effectively assembled on contact electrodes on both sides of the active layer;
- ✰ The solar cell must have a significant barrier height in the semiconductor junction;
- ✰ The impedance connected in series with the solar cell (excluding the load resistance) must be small in order to reduce power losses (Joule heat) during operation; The structure of the thin film must be uniform throughout the active region of the solar cell to avoid shorting and the effect of shunting resistances on the characteristics of the element.

The production of structures based on single-crystal silicon meeting these requirements is a technologically complex and expensive process. Therefore, attention was paid to materials such as alloys based on amorphous silicon (a-Si: H), gallium arsenide and polycrystalline semiconductors.

2. Silicon Materials

Amorphous silicon acted as a cheaper alternative to a single crystal [2]. Optical absorption of amorphous silicon is 20 times higher than that of crystalline silicon. Therefore, for a significant absorption of visible light, a-Si: H film is 0.5-1.0 μm thick instead of costly silicon 300-μm substrates. While the maximum efficiency of experimental elements based on a-Si: H - 12 per cent - slightly below the efficiency of crystalline silicon solar cells (~ 15 per cent). However, it is possible that with the development of technology, the efficiency of elements based on a-Si: H will reach a theoretical ceiling of 16 per cent. The simplest constructions of SEs from a-Si: H were created on the basis of the metal-semiconductor structure (Schottky diode) (Figure 14.1).

Despite the apparent simplicity, their implementation is quite problematic - the metal electrode must be transparent and uniform in thickness, and all states on the metal/a-Si: H boundary are stable in time. Most often, solar cells based on a-Si: H are formed on a strip of stainless steel or on glass substrates coated with a conductive layer.

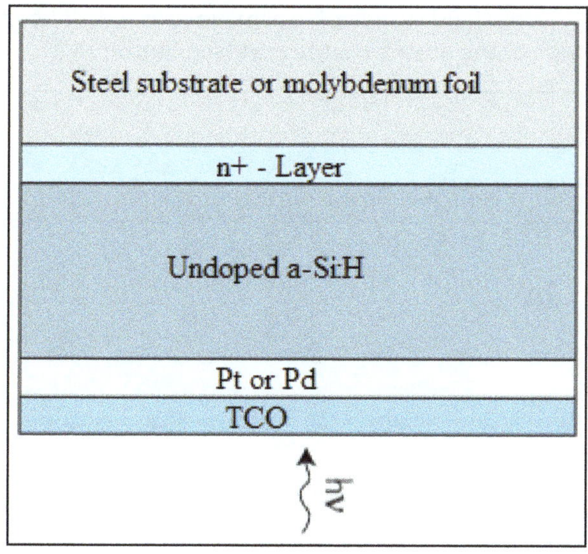

Figure 14.1: The Construction of a Photocell with a Schottky Barrier.

When glass substrates are used, a transparent conductive oxide film (TCO) of SnO_2, In_2O_3 or $SnO_2 + In_2O_3$ (ITO) is applied on them, which allows the element to be illuminated through the glass. Since the electron conductivity is weakly expressed in an undoped layer, the Schottky barrier is created by the deposition of metal films with a high work function (Pt, Rh, Pd), which causes the formation of a region of positive space charge (depleted layer) in a-Si: H. When an amorphous silicon is deposited on a metal substrate, an unwanted potential barrier a-Si: H/ metal substrate is formed, the height of which must be reduced. For this purpose, substrates of metals with a small work function (Mo Ni, Nb) are used. Before depositing amorphous silicon, it is desirable to precipitate a thin layer (10-30 nm) of a-Si: H doped with phosphorus on a metal substrate. It is not recommended to use metals that diffuse easily into amorphous silicon (for example, Au and Al), as well as Cu and Ag, as materials of electrodes, since a-Si: H has poor adhesion to them. We note that Uxx solar cells with a Schottky barrier based on a-Si: H usually do not exceed 0.6 V. The SE-based amorphous silicon with a p-i-n structure has a higher efficiency (Figure 14.2). This is the "merit" of the wide unalloyed i-region a-Si: H, which absorbs a significant fraction of the light. But the problem arises - the diffusion length of holes in a-Si: H is very small (~ 100 nm); therefore, in solar cells based on a-Si: H, charge carriers reach the electrodes mainly [3].

Only due to the internal electric field, *i.e.* Due to the drift of charge carriers. In solar cells based on crystalline semiconductors, charge carriers, having a large diffusion length (100-200 μm), reach the electrodes even in the absence of an electric field. To obtain effective SEs based on the p-i-n structure of amorphous hydrogenated silicon, it is necessary to achieve in the entire i-region a homogeneous powerful internal electric field sufficient to achieve carrier drift length commensurate with the dimensions of the absorption region (see Figure 14.2).

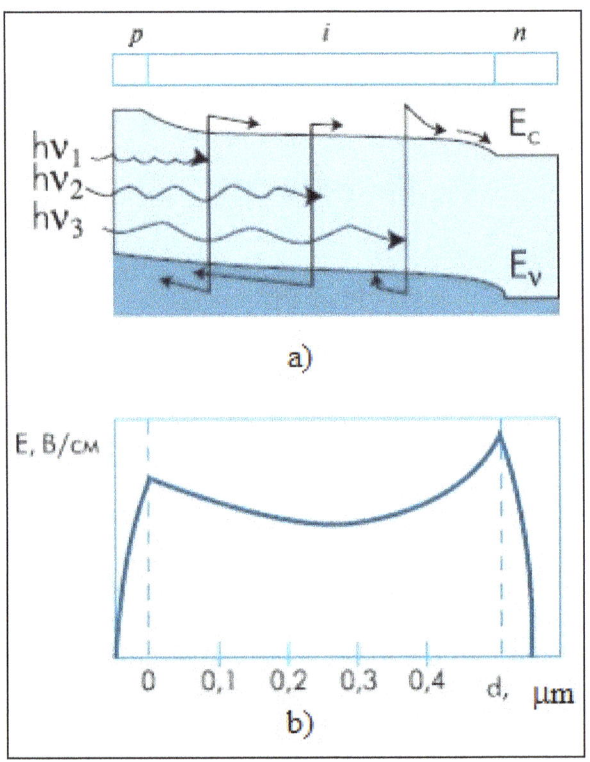

Figure 14.2: Energy Band Diagram of the p-i-n-Structures (a) and the Calculated Distribution of an Electric Field (b).

This problem is solved if the p-i-n-structure is first formed into a p-layer (Figure 14.3). To create it, a small amount of boron (<1018 cm^3) is needed, and therefore, no significant contamination of the undoped layer occurs. At the same time, if the n-layer is deposited first, the presence of residual phosphorus changes the properties of the i-layer. The formation of a p-layer on the surface of a transparent conductive electrode provides good electrical contact with it. However, the thickness of the p-layer should be small (10 nm), so that the main part of the light is absorbed in the i-region.

Another p-i-n structure of the a-Si: H cell based on a metal foil substrate, in particular stainless steel, is also used. The light comes from the side of the transparent electrode, which contacts the n-region. As a result, the short-circuit current density increases due to the reflectivity of the metallic substrate and less optical absorption of light by phosphorus-doped a-Si: H (n-region) films compared to p-doped boron layers.

One of the most promising materials for creating high-efficiency solar cells is gallium arsenide. The main advantage of gallium arsenide and alloys based on it is a wide range of possibilities for the design of solar cells [4, 5]. A GaAs-based photocell may consist of several layers of different compositions. This allows the developer to manage the generation of charge carriers with high accuracy, which in silicon AEs is limited by the allowable level of doping. A typical solar cell based on GaAs is co-based on a very thin AlGaAs layer as a window. The main disadvantage of gallium arsenide is its high cost.

3. Materials of Thin-Filk Solar Cells

The main advantage of thin-film solar cells is the saving of expensive semiconductor materials. Technologies of thin-film elements allow creating semiconductor structures on flexible substrates. At present, the main methods for obtaining films are: the glow discharge method in a mixture of silane-containing gasses, chemical deposition from the gas phase, and sputtering.

In most industrial installations, a standard plasma chemical deposition method is used at a frequency of 13.56 MHz. The essence of the method is the decomposition of silane-containing mixtures in a glow discharge plasma to the formation of active components, followed by their deposition on the growth surface. The gas containing silicon, usually monosilane SiH4 with various diluents, most often with hydrogen, is introduced into the vacuum chamber. The gas discharge is maintained by an electric field produced by a high-frequency power supply between two flat electrodes. Compared to amorphous silicon, microcrystalline silicon has a high conductivity, a high mobility of the current carriers, and a larger value of the absorption coefficient in the infrared region of the spectrum [6]. The most important role in the use of a-Si: H films for the creation of solar cells was played by the fact that the optical absorption of a-Si: H is 20 times greater than the optical absorption of crystalline silicon. For a significant absorption of visible sunlight, it is sufficient to obtain a-Si: H films 0.5-1.0 μm thick instead of using expensive silicon substrates 300 μm thick. In addition, the technological possibility of obtaining layers of amorphous silicon in the form of thin films of the large area is also promising. The advantages of solar cells based on a-Si: H in comparison with similar polycrystalline silicon elements are due to their lower manufacturing temperatures (573 K), which makes it possible to use cheap glass substrates with transparent conductive oxides (TCO) deposited on their surface, serving as electrodes of a current collector.

4. Polycrystalline Thin Films

Extremely high ability to absorb solar radiation from copper and indium diselenide ($CuInSe_2$) - 99 per cent of the light is absorbed in the first micron of this material (the width of the forbidden band is 1.0 eV) [7]. The most common material for making a solar cell window based on $CuInSe_2$ is CdS. One of the main ways of obtaining $CuInSe_2$ is electrochemical precipitation from solutions of $CuSO_4$, In2 (SO4) 3 and SeO_2 in deionized water with a Cu: In: Se ratio of 1: 5: 3 and pH of 1.2-2, 0. another promising material for photovoltaics is cadmium telluride (CdTe).

Like $CuInSe_2$, the best elements based on CdTe include a heterojunction with CdS as a window layer. Tin oxide is used as a transparent contact and an antireflection coating. A serious problem in the application of CdTe is the high resistance of the p-CdTe layer, which leads to large internal losses. But it is solved in the p-i-n-structure with the CdTe/ZnTe heterojunction (Figure 14.3). $CuGaSe_2$ is also very interesting as a thin film solar cell element.

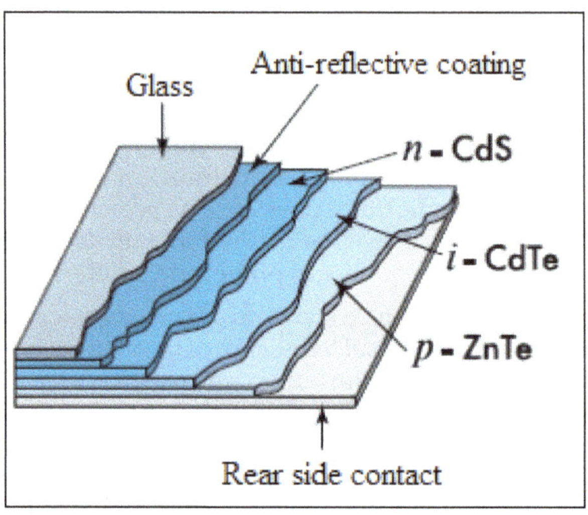

Figure 14.3: Structure of a Solar Cell based on CdTe.

One of the promising materials for cheap solar cells due to the acceptable bandgap width (1.4-1.5 eV) and a large absorption coefficient of 104 cm-1 is Cu2ZnSnS4.

Among the solar cells, a special place is occupied by batteries using organic materials [5]. In particular, the efficiency of a solar cell based on titanium dioxide coated with an organic dye is very high ~ 11 per cent. It is also important that polymer films can be used as substrates in such elements. The basis of this type of solar cell is a wide-band semiconductor, usually, TiO2, coated with a monolayer of organic dye. The photoelectrode of such a device is a nanoporous TiO2 film 1 μ thick deposited on the TCO on the glass. The reflecting electrode is a thin layer of Pt deposited on the TCO on the glass.

The principle of operation of the element is based on the photoexcitation of a dye and the rapid injection of an electron into the conduction band of TiO2. In this case, the dye molecule is oxidized, electric current flows through the element and the triiodide are reduced to the iodide on the platinum electrode. Then the iodide passes through the electrolyte to the photoelectrode, where it restores the oxidized dye.

Fullerenes (C60) are also very promising for organic solar cells based on C60/p-Si heterostructures due to their ability to strongly absorb in the short-wavelength region of the solar spectrum [7]. Thermo-photovoltaic electricity production, *i.e.*, conversion of long-wavelength (thermal) radiation by means of photovoltaic cells was discovered in 1960 and is of increasing interest, especially in connection with current advances in the development of narrow-band semiconductors. In a thermophotovoltaic cell, heat is converted into electricity by means of selective emitters from rare earth oxides - erbium and ytterbium. These substances absorb infrared radiation and again emit it in a narrow energy range. Radiation can be effectively converted by means of a photovoltaic cell with an appropriate band gap. As the material for the photovoltaic cell, In Ga 1-x As is most suitable, since it allows to achieve the necessary width of the forbidden band.

5. Solar Cells Based on Heterostructures

The creation of solar cells based on AlGaAs-GaAs heterostructures opened a new page in solar photovoltaics [8]. One of the results of studies of heterojunctions was the practical implementation of the idea of a wide-gap window for solar photovoltaics. In heterostructures (AlGaAs (wide-gap window)) - (p-n-GaAs (photoactive region)), it was possible to form a defect-free heterointerface and provide ideal conditions for the photogeneration of electron-hole pairs and their p-n-junction collection. Since hetero-photo elements with an arsenide-gallium photoactive region turned out to be even more radiation-resistant, they quickly found application in space technology, despite the much higher cost compared with silicon photocells.

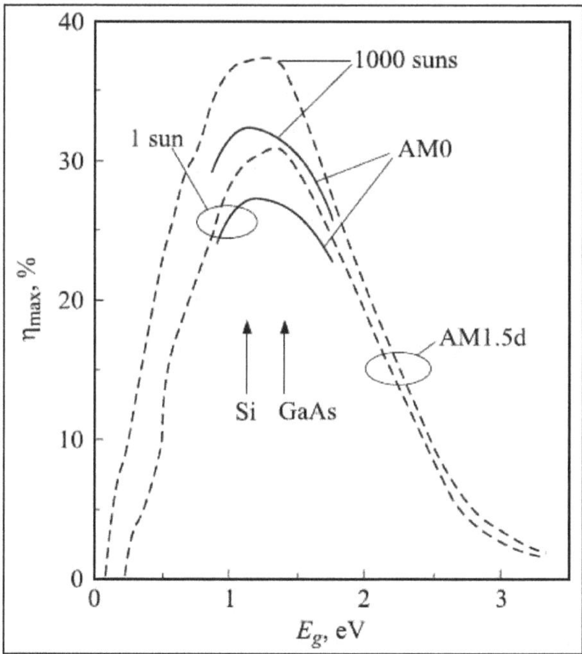

Figure 14.4: Dependences of the Maximum Achievable Conversion Efficiency (ηmax) of a Solar Cell with one p-n-junction from the Width of the Band Gap of the Material (Eg). The solid lines are for the AM0 solar spectrum, the dotted lines for the AM1.5d spectrum (for non-concentrated solar radiation (1 sun) and for 1000-fold concentrated radiation).

Silicon and gallium arsenide largely satisfy the conditions of "ideal" semiconductor materials. If we compare these materials from the point of view of the efficiency of the photoelectric conversion, they are almost identical, close to the absolute maximum for a single photocell (Figure 14.4).

For silicon photocells, a planar structure with a small p-n junction obtained by the diffusion method was used. It was necessary to use epitaxial methods for gallium arsenide-based photocells when growing the wide-gap AlGaAs window. A comparatively simple method of liquid-phase epitaxy developed earlier for obtaining the structures of the first-generation hetero- lasers was used. In the case of photocells, it was necessary to grow only one wide-gap p-AlGaAs layer, while the p-n junction was obtained by diffusion of the p-type impurity from the melt into the base material of n-GaAs (Figure 14.5a).

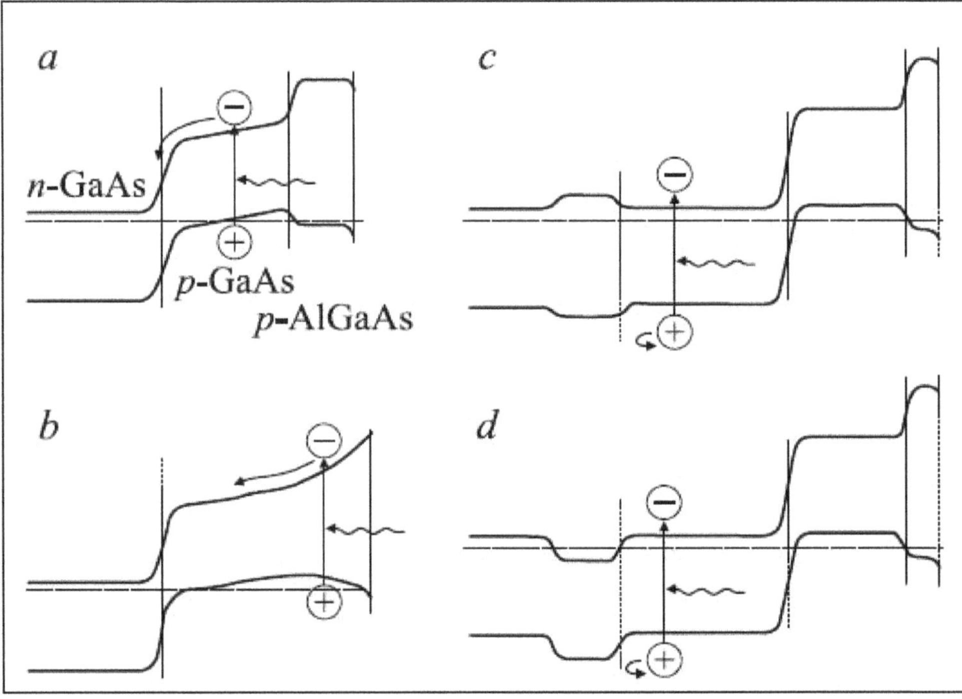

Figure 14.5: Zones Diagrams of p-AlGaAs-p-n-GaAs Heterojunction Solar Cells: a is a structure in which a p-GaAs layer with an integrated electric field is obtained by diffusing zinc into an n-GaAs base during the growth of the wide-band p-AlGaAs layer; B - structure with a strong built-in electric field; C - a structure with a rear wide-gap layer that creates a potential barrier; D is a structure with a rear potential barrier formed by a high-density n + -GaAs layer.

At the same time, progress in the field of solar photovoltaics based on gallium arsenide was due to the use of new epitaxial methods for growing heterostructures-mainly a gas-phase epitaxy from organometallic vapor (MOS GFE). This method was developed in the process of improving injection lasers and second-generation photovoltaics based on AIIIBV compounds.

What improvements have been made to the structure of solar heterophobes due to the new technological possibilities that have been discovered? First, the wide-gap AlGaAs window was optimized, the thickness of which became comparable with the thickness of nanoscale active regions in hetero-lasers. The AlGaAs layer also became a third component in the three-layer interference antireflection coating of the photocell. Along the wide-gap AlGaAs layer, a narrow-band, heavily doped contact layer is grown, which is removed during post-processing in the intervals between the contact strips. Secondly, a rear (behind the p-n-junction) wide-gap layer was introduced, providing, together with the frontal wide-gap layer, a two-sided restriction of the photogenerated carriers within the light absorption region (Figure 14.5c).

In the structures of AlGaAs/GaAs photocells grown by the MOSFET method, a single wide-gap AlGaAs layer forming the rear potential barrier could be replaced by a system of alternating pairs of AlAs/GaAs layers forming a Bragg mirror. The wavelength of the maximum in the reflection spectrum of such a mirror was chosen near the absorption edge of the photoactive region, so long-wavelength radiation that was not absorbed in this region in one pass could be absorbed in the second pass after reflection from the mirror [9]. At the same time, wide-gap mirror layers continued to function as a back barrier for photogenerated carriers. Under these conditions, the thickness of the photoactive region could be reduced by a factor of 2 without loss of current in comparison with structures without a mirror. This significantly increased the radiation resistance of photocells, since a number of defects introduced by irradiation with high-energy particles that affect the diffusion of carrier diffusion lengths decreased in proportion to the decrease in the thickness of the photoactive region [9].

Using the nonequilibrium conditions of epitaxy and (or) integrating intermediate superlattices, it was possible to find the growth conditions for perfect AlGaAs/GaAs heterostructures on a germanium substrate. Since that time, hetero-photo elements on germanium have been considered as the main candidates for use on most space vehicles. The decisive role here was played by the fact that germanium is stronger mechanically than gallium arsenide, used before as a substrate. Therefore, the batteries made up of AlGaAs/GaAs-photocells on germanium, by weight and strength characteristics, were comparable to silicon ones, and they were superior in efficiency and radiation resistance. Another "photoelectric" problem was of fundamental importance for solar photo-electro energetics.

6. Types of Organic Photocells

The most interesting are solar photocells based on semiconductor conjugated polymers [10], which can be flexible and produced using inexpensive technologies developed in the polymer industry. To realize a bulk heterojunction, a donor-acceptor polymer (donor) composite with an acceptor material is needed. As the latter, the derivatives of fullerenes, polymeric and low-molecular acceptors are most often used.

Polymer-Fullerene Photocells

Polymer-fullerene SFE based on bulk heterojunction has been shown that the addition of C60 fullerene to the conjugate polymer poly-methoxy- (2-ethyl hexyl oxy) -1,4-phenylenevinylene (MEH-PPV) increases the efficiency of PV by several orders of magnitude. A significant fraction of the fullerene in the composite is necessary to achieve a balance of mobilities of electrons and holes in a bulk heterojunction. In recent years, the greatest attention has been attracted to SFEs based on a bulk heterojunction from poly [3-hexylthiophene] (P3HT) and PCBM with an efficiency of 4-5 per cent. We note that the photogeneration of electrons and holes during photoexcitation of the active layer and their transport to the electrodes in the bulk heterojunction P3HT/PCBM are performed with very high efficiency. Thus, the external quantum efficiency of SFE, *i.e.* the number of charge carriers collected on electrodes, counting on the incident photon, reaches 70 per cent. The dissociation efficiency of the bound electron-hole pair (exciton) to the free one reaches 90 per cent. Such high figures were obtained by optimizing the morphology of the bulk heterojunction P3HT/PCBM and achieving the phase separation of the donor (P3HT) and acceptor (PCBM) with a characteristic scale of tens of nanometers [11].

The main possibilities for increasing the efficiency of polymer-fullerene SFEs are associated with the expansion of the absorption spectrum of the donor-acceptor nanocomposite to increase the short-circuit current (Ik) and with the decrease in the ED (LMO) -EV (EA) energy difference for increasing the idling voltage (Vxx).

The monomer unit of one of the most successful narrow-gap PCPDTBT polymers for solar photocells includes donor and acceptor fragments, which made it possible to achieve an optical gap of 1.46 eV (850 nm absorption edge). SFE based on the PCPDTBT nanocomposite with methanol-fullerene PCBM [C70] showed a record efficiency of 5.5 per cent today [12].

Noticeable voltage losses of polymer-fullerene SFEs are due to the strong inequality ED (HBMO) -EA (HBMO)> Eb. In one of the most successful three-dimensional heterojunctions P3HT/PCBM ED (HBMO) -EA (HBMO) 1 eV, while Eb is estimated at 0.3-0.4 eV (Figure 14.5b). To reduce the difference between ED (HBMO) - EA (HBMO) and thereby increase Vxx can be achieved by developing conjugated polymers with a higher ionization potential: in [13] it is shown that Vxx is linearly dependent on ED (HOMO).

In addition, fullerene metal complexes have a stronger optical absorption in the red spectral range than organic fullerene derivatives, which can be used to increase the photocurrent of polymer-fullerene SFE.

One of the drawbacks of fullerene C60 and its organic derivatives is low absorption in the visible region of the spectrum, which limits the absorption of the polymer-fullerene composite. At the same time, higher fullerenes, for example, C70, have a much higher absorption. Thus, the most effective polymer-fullerene SFE used metallofullerene PCBM [C70] [12].

Attempts are being made to find more suitable acceptors for polymer SFE. As such acceptors, perylene derivatives, nanotubes, fullerene dimers and oligomers and other substances and nanoparticles are studied. One of the ways to overcome these difficulties can be the use of intermolecular complexes with charge transfer formed between the conjugated polymer and a suitable acceptor [14]. Such complexes based on the model conjugated polymer MEH-PPV can effectively absorb the red part of the solar spectrum, generate free charges and be used to optimize the morphology of the donor-acceptor composite [47]. Finally, polymer complexes with charge transfer showed an exceptionally high photostability [15].

Polymer Photocells

The idea of a polymer-polymer bulk heterojunction looks attractive because, with the help of two suitable conjugated polymers, it is possible to make a composite with an absorption corresponding to the spectrum of the Sun with an optimal difference in the energies of the boundary orbitals. The main difficulties in the field of polymer SFEs are related to the development of an acceptor polymer with sufficient electron mobility and to optimize the nanomorphology of the donor-acceptor composite.

Anodium Copper Oxide

In favor of the choice of copper as a starting material, the following considerations speak. Compared to copper, modern materials for efficient photoconversion (Si, GaAs) have a higher cost. The Cu_2O film is formed by a low-cost method of anodic oxidation. Semiconductor solar cells are sensitive to temperature increase, which is why the effectiveness of using focusing systems is reduced, since the efficiency of photoconversion decreases when the semiconductor overheats.

Methods for the electrochemical formation of a copper-based oxide film in aqueous solutions of sulfuric acid, hydrochloric acid, in an alkaline solution are known. The main problem of these methods is the presence of two oxides (Cu_2O and CuO) in the film composition, whereas the photoactive component of the film is precisely the monovalent oxide Copper [16].

Analyzing the literature and experimental data, it can be stated that Cu_2O films obtained by low-temperature, low-cost anodic oxidation from affordable, inexpensive raw materials (electrical copper) can form the basis for the creation of solar batteries for the needs of small-scale power generation. An increase in the efficiency of photoconversion can be achieved by improving the technology (optimizing the parameters of the anode process, selecting materials and methods for creating collecting contacts, *etc.*).

7. Conclusion

New highly stable donor and acceptor materials with high mobility of charge carriers and high optical absorption are in the visible spectral range. Such materials should form a volumetric heterojunction with a characteristic scale of the donor and acceptor phase separation into tens of nanometers at the optimum energy difference of the boundary molecular orbitals of donor and acceptor materials. Due to the relatively low cost of production of polymer SEs in comparison with solar cells based on inorganic semiconductors, the use of solar cells based on organic and hybrid materials has obvious advantages. To date, the efficiency of energy conversion in polymer tandem solar cells, obtained in laboratory conditions, reaches ~ 9 per cent, which is already approaching the commercially competitive values of 10-11 per cent. Significant progress in the field of synthesis of new polymer materials for solar cells, as well as in the field of production and research of the properties of modern polymer SEs, left no doubt in recent years that this level of efficiency of polymer SE will be achieved in the near future.

Solar photoelectric power is not born from scratch. Largely due to the development of electronics, laser technology, and electric power engineering for space vehicles, a scientific and technological base has been created that can serve as a starting point for the deployment of ground-based solar power engineering based on semiconductors. There comes a time when it is necessary to move to a wider investment of funds in this area, corresponding to the significance that the solar power industry will have in the future.

The concept of bulk heterojunction proved to be the most effective approach for the development of organic and hybrid SFE with high efficiency. The main task now is to increase the efficiency and lifetime of solar photocells, which requires new highly stable donor and acceptor materials with the mobility of charges of at least 10-4 cm2/V s and strong optical absorption in the visible spectral range. The development of new and optimization of existing photocatalysts active in the visible and near-infrared regions of the spectrum will allow the creation of new efficient water and air purification systems contaminated with toxic organic compounds and various microorganisms.

REFERENCES

1. Ryazanov K.V. Perspektivy razvitia solneshnoi energetiki//KABEL–news. – 2009. – 12–1. – S. 81–85.
2. Saga T. Advances in crystalline silicon solar cell technology for industrial mass production//npg asia materials. – 2010. – . 2. – . 3. – Ñ. 96-102.
3. Purification of silicon for photovoltaic applications. Ayes Delannoy, Journal of Crystal Growth 360, (2012), ð. 61-67
4. Processes for upgrading metallurgical grade silicon to solar grade silicon. J. Safarian, G. Tranell, M.Tangstad. Energy Procedia 20, 2012, pp. 88-97
5. Elkem Solar siliconTM – staying relevant – innovation for the oversupplied market. Torgeir Ulset. TeknovaSolar Energy Conference. 11.06.2013, Kristiansand
6. Kazansky A.G. Photoelectricheskie svoistva mikrokristallicheskogo kremnia// Izv. vuzov. Ser. Materiali. electronnoi tehniki. – 2009. – 1. –p. 12–21.
7. Light S., Khaselev O., Ramakrishna P.A., Faiman D., Katz E.A., Shames A., Goren S. Fullerene Photoelectrochemical Solar Cells. – Solar Energy Materials and Solar Cells, 51(1998), p.9–19.
8. M.E. Green, K. Emery, D.L. King, S. Igary, W. Warta. Progr. Photovolt.: Res. Appl., 10, 355 (2002).

9. V.M. Andreev, I.V. Kochnev, V.M. Lantratov, M.Z. Shvarts. Proc. 2nd World Conf. on Photovolt. Solar Energy Conversion (Vienna, 1998) p. 3757.

10. Li G., Shrotriya V., Huang J.S., Yao Y., Moriarty T., Emery K., Yang Y. Nature Materials, 2005, v. 4, 11, p. 864–868.

11. Moule A.J., Meerholz K. Adv. Mater., 2008, v. 20, p. 240–245.

12. Peet J., Kim J.Y., Coates N.E., Ma W.L., Moses D., Heeger A.J., Bazan G.C. Nature Materials, 2007, v. 6, 7, p. 497–500.

13. Scharber M.C., Muhlbacher D., Koppe M., Denk P., Waldauf C., Heeger A.J., Brabec C.L. Ibid., 2006, v. 18, 6, p. 789– 794.

14. Bruevich VV, Makhmutov T.Sh., Elizarov SG, Nechvolodova EM, Parashchuk D.Yu. G. experiment. Theor. Physics, 2007, v. 132, No. 3, p. 531-542.

15. Golovnin I.V., Bakulin A.A., Zapunidy S.A., Nechvolodova E.M., Paraschuk D.Yu. Appl. Phys. Lett., 2008, v. 92, 24, p. 243311–3.

16. Ustinov, VI, S.A. Podorozhnyak, G.N. Shelovanov. Investigation of anodic copper oxide/V.I. Ustinov//Materials of the 50th jubilee international scientific student conference, section: Chemistry//NSU - Novosibirsk 2012.-p. 190-191.

Chapter 15

The Solar Resource

Ahmed H. Mmingwa

Tanesco, Ubungo Umeme Park Building,
Dar Es Salaam, Tanzania
E-mail: ahamed.mmingwa@tanesco.co.tzm mmingwa2002@yahoo.com

The purpose here is to explain how the solar PV system works and sunlight being fundamental in function of portion electromagnetic radiation given off by the sun in particular infrared, visible, and ultraviolet light on Earth sunlight is filtered through Earth's atmosphere and is obvious as daylight when the sun is above the horizon.

Power inverters which aim to efficiently transform a DC power source to a high voltage AC source similar to power that would be available at an electrical wall outlet, inverters are used for many applications as in situations where low voltage DC sources such as batteries, solar panels or fuel cells must be converted so that devices can run off AC power one example of such a situation would be converting electrical power from a car battery to run a laptop, television set or a cell phone.

The method in which low voltage DC power is inverted is completed in two steps, the first being the conversion of the low voltage DC power to a high voltage DC source, and the second step being the conversion of the high DC source to an AC power waveform using pulse width modulation.

Another method to complete the desired outcome would be to first convert the low voltage DC power to AC power and the use a transformer to boost the voltage to say 120/220 volts, of the different DC-AC power inverters on the market today there are essentially two different forms of AC power output generated, modified sine wave, and pure sine wave.

A modified sine wave can be seen as more of a square wave than a sine wave it passes the high DC voltage for specified amounts of time so that the average power and rms voltage are the same as if it were a sine wave, these types of inverters are much cheaper than pure sine wave inverters and therefore are attractive alternatives.

1. Introduction

Sunlight is a portion of the electromagnetic radiation given off by the sun in particular infrared, visible, and ultraviolet light. On Earth sunlight is filtered through Earth's atmosphere, and is obvious as daylight when the Sun is above the horizon. When the direct solar radiation is not blocked by clouds it is experienced as Sunshine a combination of bright light and radiant heat.

When it is blocked by clouds or reflects off other objects it is experienced as diffused light the World Meteorological Organization uses the term "Sunshine duration" to mean the cumulative time during which an area receives direct irradiance from the Sun of at least 120 watts per square meter. Other sources indicate an Average over the entire earth of 164 Watts per square meter over a 24 hour day.

The ultraviolet radiation in sunlight has both positive and negative health effects as it is both a principal source of vitamin D_3 and a mutagen sunlight takes about 8.3 minutes to reach the Earth from the surface of the Sun and sunlight is a key factor in photosynthesis, the process used by plants and other autotrophic organisms to convert light energy normally from the sun, into chemical energy that can be used to fuel the organisms' activities.

Ultraviolet C or (UVC) range, which spans a range of 100 to 28 nm the term ultraviolet refers to the fact that the radiation is at higher frequency than violet light (and hence also invisible to the human eye)due to absorption by the atmosphere very little reaches Earth's surface this spectrum of radiation has germicidal properties as used in germicidal lamps.

Ultraviolet B or (UVB) range, spans 280 to 315 nm it is also greatly absorbed by Earth's atmosphere and along with UVC causes the photochemical reaction leading to production of the ozone layer. It directly damages DNA and causes sunburn, but is also required for vitamin D synthesis in the skin and fur of mammals.

Ultraviolet A or (UVA) spans 315 to 400 nm this band was once held to be less damaging to DNA and is used in cosmetic artificial sun tanning (tanning booths and tanning beds) and PUVA therapy for psoriasis, however UVA is now known to cause significant damage to DNA via indirect routes (formation of free radicals and reactive oxygen species) and can cause cancer.

Visible range or light spans 380 to 780 nm as the name suggest this range is visible to the naked eye it is also the strongest output range of the sun's total irradiance spectrum.

Infrared range that spans 700 nm to 1,000,000 nm (1 mm) it comprises an important part of the electromagnetic radiation that reaches Earth. Scientists divided the infrared range into three types on the basis of wavelength as follows:-

- ☆ Infrared – A 700nm to 1,400 nm
- ☆ Infrared – B 1,44 nm to 3,000 nm
- ☆ Infrared – C 3,000 nm to 1 mm

Researchers may record sunlight using sunshine recorder to calculate the amount of sunlight reaching the ground, both Earth's elliptical orbit and the attenuation by Earth's atmosphere have to be taken into account, the extraterrestrial solar luminance (E_{ext}), corrected for the elliptical orbit by using the day number of the year.

2. Materials and Methods

Solar Power Kits as know by many include solar panels, inverters, cables, mounting structures and batteries (hybrid kits only) vary widely in their power producing capacity, solar energy-Power from the sun is a vast, inexhaustible and clean resource for electricity generation represents a clean alternative to electricity from the fossil fuels with no air and water pollution, no global warming, no risk of electricity price spikes and no threats of public health, solar energy can heat water, cool and heat homes, provide natural lighting and when a system is set in place can convert the solar resource into a useful energy and the fuel is free.

Solar energy in fact is not cheap one could argue that from a cost savings point of view it is not very practical at all because it typically will take many, many years to reach the break-even point when considering the cost of your local utility electricity.... I'm talking roughly 5 to 15 years in many instances depending on your usage.

However, despite the cost of solar power systems for many folks it is a worthwhile investment for reasons other than saving money on your utility bill, if your property is far from the nearest road it actually cost less to have to have solar power than to pay to run electricity to your property.

The World Meteorological Organization (WMO) uses the term ''sunshine duration'' to mean the cumulative time during which an area receives direct irradiance from the sun of at least 120 watts per square meter while sources indicate an average over the entire earth of 164 watts per square meter over a 24 hour day.

The total amount of energy received at ground level from the sun at the zenith depends on the distance to the sun and thus on the time of year, it about 3.3 per cent higher than average in January and 3.3 per cent lower in July, if the extraterrestrial solar radiation is 1367 watts per square meter, then the direct sunlight at earth's surface when the sun is at the zenith is about 1050 watts per square meter.

Connection Methods

As has been indicated in Figure 15.1.

3. Results and Discussion

In the world today there are currently two forms of electrical transmission Direct Current (DC) and Alternating Current (AC) each its own advantages and disadvantages. DC power is simply the application of a steady constant voltage across a circuit resulting in a constant current.

A battery is the most common source of DC transmission line as the current flows from one end of the circuit to the other most digital circuitry of today is run off of DC power as it carries the ability to provide either a constant high or a constant low voltage enabling digital logic to process code executions.

Alternating Current (AC power) unlike Direct Current (DC power) oscillates between two voltages value at specified frequency (60/50 Hz) its ever changing current and voltage makes it easy to step up the voltage for high voltage long

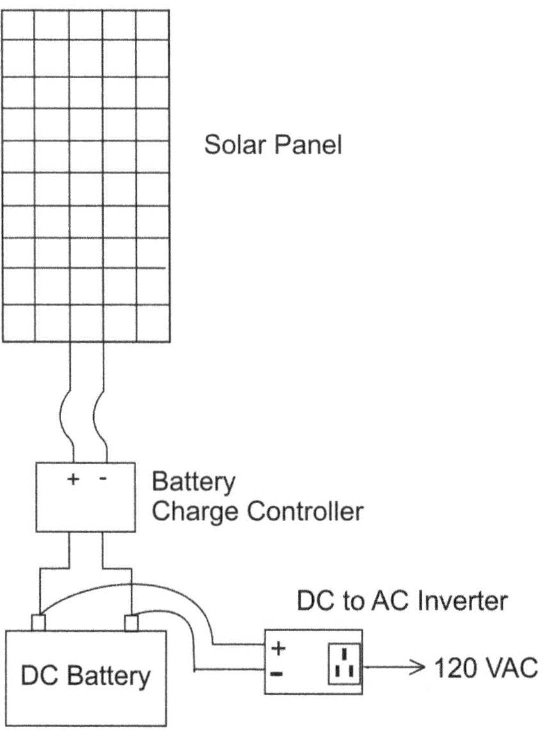

Figure 15.1

distance transmission line and step down the voltage to an appropriate voltage value through(up/down) transformers.

Solar panels work through what is called a photovoltaic process where radiation energy (photo) is absorbed and generate electricity (voltaic) radiation energy is absorbed by semi conductor cells normally silicon material and transformed from photo energy (light) into voltaic (electric current).

When the sun's radiation hits a silicon atom a photon of light is absorbed knocking off an electron these released electron create an electric current, the electric goes to an inverter which convert the current from direct current (DC) to an alternating current (AC). The electrons flow out of these solar and into an inverter and other safety devices the inverter converts the direct current (DC power commonly used in batteries) into an alternating current or (AC power).

AC power is the kind of electric of electrical that our television, computers, and toasters use when plugged into wall outlet switches, solar cells are generally small, squired shaped panels semiconductors made from silicon and other conductive materials, manufactured in thin layers such that when sunlight strike a solar cells chemical reaction release electrons and generate electric current.

Power inverters are devices which can convert electrical energy of Direct Current (DC power) into that of Alternating Current (AC power), they can come in all shapes and size, from low power functions as powering a car radio that of backing up a complex building in case of power grid outage.

Inverters can come in many different varieties, different in price, power, efficient and purpose the purpose of DC/AC power inverters is typically to take DC power source, say supplied by a battery such as 12 volt car battery and transform it into a 120/220 volt AC power operating at 60/50 Hz emulating the power available at an ordinary household socket outlets for powering appliances like Radio, Televisions, computers and many more

A photovoltaic PV system needs unobstructed access to the sun's rays for most or all the day to be effective shading on the system can significantly reduce energy output while climate change is not a major concerned because PV system are relatively unaffected by air temperature, abundant year round sunshine make solar energy system useful and effective.

The size of your solar system depends on several factors such as how much electricity, hot water, space heating you use, the size of your roof, how much you are willing to invest and how much you want to generate. It is much better to contact a system designer/installer to determine for you what type of PV system fits your requirements.

4. Policy and Plants of the Government of Tanzania

In collaboration with solar agencies in the country under the Ministry of Energy and Minerals (MEM) for the year 2014/2015 installed 14 small solar system for off Grid villages in the following Districts namely Mlele, Katavi, Myui, Namanyere, Nkasi, Ngara, Biharamulo, Chato, Ukerewe, Mpanda Kalihuwa, Bigiri, isaka, Bwawani and Kongwa

However solar systems to be connected into National Grid is currently understudy in Kishapu Shinyanaga and Dodoma Regions for 150 MW, 50 MW by the year 2020 and 2019 respectively

5. Conclusions

In most of African countries sun rays for PV system are available almost in 24 hours a day, arrangement need be in place and establishing solar working stations in particular areas where national grid network not available.

As mentioned in here for small PV system in village areas the cost of it is relative low of which each respective African countries can afford to set in their budget to save majorities of people in the villages

The application of solar power panels can be used in different ways and purpose such as in big cities for outage of grid system, big towns, as well as in villages where there is no grid connections as here under:

- ✰ *Figure 15.2 is small PV system installed at Mlalo Village in Muheza District*
- ✰ *Figure 15.3 is in Powaga irrigation system in Rural Iringa District*

Figure 15.2: Small PV System Installed at Mlalo Village in Muheza District.

Figure 15.3: Powaga irrigation System in Rural Iringa District.

REFERENCES

1. Bird L.,J.McLaren,J.Heeter,C.Linvill,J.Shenot, R.Sedano,and J.Migden-Ostrander. 2013. Regulatory considerations associated with the expanded adoption of distributed solar. Golden,CO:National Renewable Energy Laboratory.
2. International Energy Agency (IEA) 2014 – Technology Roadmap: Concentrating Solar Power.Paris, France.
3. Burger, B.2011. Solar power plants deliver peak load.Friburg,Germany:Fraunhofer Institute for Solar Energy System ISE
4. Fraunhofer Institute. 2015 Photovoltaic report.

Chapter 16

Simulation and Prediction of the Power Output and the Photocurrent for Photovoltaic Systems

K. Kety, A.K. Amou, K. Sagna, Y. Lare and K. Napo

Laboratoire sur l'Energie Solaire,
Université de Lomé, Togo
E-mail: mapkamou@yahoo.fr

We study in this paper, based on comparison already made in the literature concerning photovoltaic generator power models, the most optimal model applied to the operation of the photovoltaic generator of Sévagan (Togo). The comparison with the experimental data is carried out, which allowed us to verify the validity of the model. Finally, the influence of the characteristic parameters on the photovoltaic module ECO LINE LX-260P used to make the photovoltaic generator of the Sévagan dispensary (in Togo) is studied in order to predict the power production of the module according to the meteorological conditions (temperature-Irradiation). The comparison with the experimental data will be carried out in order to verify the validity of the model. To verify the validity of the model throughout the range of weather conditions, the process was done in two steps: on a sunny day and a cloudy day. A good agreement was observed with 95 per cent, 97 per cent and 99 per cent correlation coefficients for cloudy, sunny days and the generator photocurrent simulation respectively. The results demonstrate an acceptable accuracy of the power model under different environmental conditions

***Keywords**: Simulation, Photovoltaic generator, Power output models - Prediction, Optimization.*

1. Introduction

The design of optimized photovoltaic systems is by nature difficult. Indeed, as far as the source of a photovoltaic system is concerned, the power produced change greatly as a function of the irradiation, the temperature, but also the global aging of the system. These variables are influencing the behavior of the system, according to daily and seasonal fluctuations. For these reasons, the photovoltaic panel can provide maximum power only for a particular voltage and a certain current. This operation at maximum power depends on the load at its terminals whether of continuous or alternative nature. In order for the generator to work most often in its optimum speed, the commonly adopted solution is to introduce a static converter which will act as a source-charge adapter: Maximum Power Point Tracker (MPPT). For given conditions, the peak power of the installed photovoltaic generator is best exploited at the maximum power point of the power characteristic as a function of the voltage. These points therefore correspond to the point of optimum power, the term reflecting the character relating to the illumination and temperature conditions of the power supplied. In the literature, there are many simplified mathematical models to determine the maximum power supplied by a photovoltaic generator as a function of variations in solar irradiation and ambient temperature. For example, (Borowy and Salameh, 1996) gave a simplified model whereby the maximum power produced can be calculated for a certain photovoltaic module after the solar irradiation on the photovoltaic module and the temperature are determined. Other researchers (Jones and Underwood, 2002) proposed another model to calculate the maximum electrical power at the terminals of a photovoltaic sensor. The latter has a reciprocal relationship with the temperature of the module and it has in addition a logarithmic relationship with the solar irradiation absorbed by the module. (Lu Lin, 2004) developed another model to calculate the maximum power supplied by a photovoltaic module for a given sunshine and module temperature with four constant parameters to be determined experimentally. A presentation of the different power models is made in this paper. On the basis of the comparisons already made in the literature, the most optimal model will be chosen for application to the Sévagan photovoltaic generator. The comparison with the experimental data will be carried out in order to verify the validity of the model.

2. Materials and Methods

2.1. Materials and Experimental Method

The experimental equipment consists of an AHLBORN pyranometer type FL A613-GS, SN: 15111835/15, a thermometer serving as a temperature sensor, an ALMEMO dataloggerand multimeters for measuring the current and voltage of ECO LINE LX-260P solar module.

Measurements were carried out at the Solar Energy Laboratory of the University of Lomé (LES-UL) on the ECO LINE LX-260P module as well as solar irradiation and ambient temperature. The same mesurements were all so done in Sévagan for the photovoltaic generator constitued by 11 modules ECO LINE LX-260P. All these measurements were made with an accuracy of 10^{-4}. These data were automatically recorded using the ALMEMO control unit.

2.2. Power Models

2.2.1. Model 1: Benchmark Model (Lu Lin, 2004)

This model was developed and validated experimentally by (Lu Lin, 2004). It allows to determine the maximum power supplied by a photovoltaic module, for a sunshine G and a given module temperature with only four constant positive parameters, to be determined. The parameters a, b, c and d can be known experimentally, and a system of simple equations can be solved resulting in a sufficiently extended set of measurement points (Belhadj, 2010 and Jones, 2002).

$$P_{max,1} = -(aG + b).T_c + c.G + d \quad (1)$$

with $P_{max,1}$ the maximum power produced in Watt;

T_c: the temperature of the module that can be described by Equation 2 According to the experimental measurements carried out on a module (BP Solar 340), the constants a, b, c and d are 0.0002; 0.0004; 0.1007 and 0:1018 respectively. The NOCT of the module BP Solar 340 is 47∓2 C; which allows us to find:

$$T_c = T_a + 0{,}03375 \quad G \quad (2)$$

So equation [1] becomes:

$$P_{max,1} = -(a.G + b).(T_a + 0{,}03375 \quad G) + c.G + d \quad (3)$$

2.2.2. Model 2: Input/output power model

The power produced by a photovoltaic generator is estimated from the data of the global irradiation on an inclined plane, the ambient temperature and the data of the manufacturer for the photovoltaic module used. It is given by equation (4): (Fethi, 2015 ; R. Zeiba, 2016)

$$P_{max,2} = \eta_g.S.N.G. \quad (4)$$

With S: area of the module constituting the photovoltaic field,

N: Number of photovoltaic field modules, G: Irradiation on inclined plane,

η_g: Instantaneous efficiency of the photovoltaic generator given by equation [5]:

$$\eta_g = \eta_r.\eta_{pt}.[1 - \beta_t(T_c - T_r)] \quad (5)$$

η_r: Reference efficiency of the modules under standard conditions,

η_{pt}: Efficiency of the tracking system of the maximum power point which is equal to 1 if a perfect system is used,

T_r: Reference temperature,

β_t: Experimentally determined temperature coefficient of yield.

It is defined as the variation of the efficiency of the module for a 1 C variation in the temperature of the cell. Typical values for this coefficient are between 0.004 and 0.006 (Fethi, 2015; R. Zeiba, 2016).

2.2.3. Model 3: Simulation Model of Borowy and Salameh

This model was developed by (Borowy and Salameh, 1996). The principle of the model is based on the equivalent circuit with one diode (Figure 16.1).

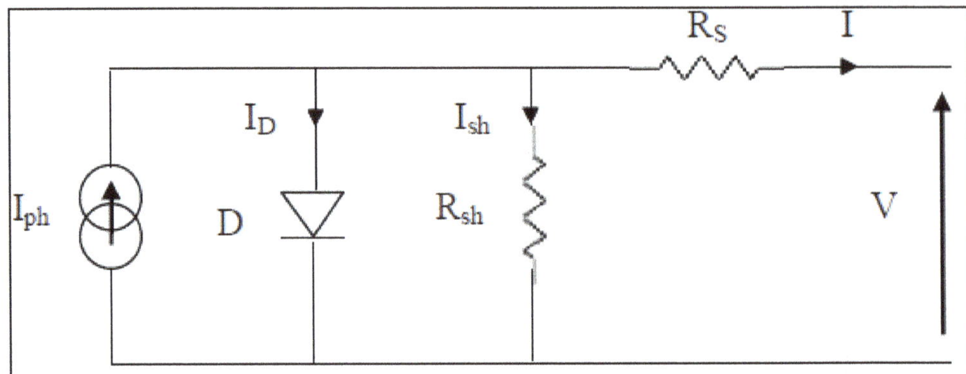

Figure 16.1: Equivalent Diagram of a Solar Cell: Model with One Diode.

This model use the characteristics of the modules provided by the manufacturers. It therefore offers a very simple way of calculating the power produced by the PV modules. The formulas for calculating the maximum current intensity as well as the maximum voltage of the module under arbitrary conditions are given by the following relations (Lu Lin, 2010; Belhadj, 2008):

$$I_{max,3} = I_{cc} \cdot \left\{1 - C_1 \left[\exp\left(\frac{V_m}{C_2 \times V_{co}}\right) - 1\right]\right\} + \Delta I \qquad (6)$$

I_{cc}: Short-circuit current of the module (in Ampere),

V_{cc}: Open circuit voltage of the module (in Volt).

The parameters C_1 and C_2 are calculated by equations [7] and [8]:

$$C_1 = \left(1 - \frac{I_m}{I_{cc}}\right) . \exp\left(-\frac{V_m}{C_2 . V_{co}}\right) \qquad (7)$$

$$C_2 = \frac{\frac{V_m}{V_{co}} - 1}{\ln\left(1 - \frac{I_m}{I_{cc}}\right)} \qquad (8)$$

I_m: Maximum current of the module under standard conditions,

V_m: Maximum voltage under standard conditions

The maximum voltage under arbitrary conditions is [3]:

$$V_{max,3} = V_m \cdot \left[1 + 0{,}0539 . \log\left(\frac{G}{G_0}\right)\right] + K_V . \Delta T - R_s . \Delta I \qquad (9)$$

G_0: Solar irradiation under standard conditions ($\frac{1000W}{m^2}$),

K_V: Temperature coefficient of open circuit voltage,

R_s: Series resistance of the module.

$$\Delta T = T_c - T_r \tag{10}$$

$$\Delta I = K_I.\left(\frac{G}{G_O}\right).\Delta T + \left(\frac{G}{G_O} - 1\right).I_{cc} \tag{11}$$

K_I is the temperature coefficient of the short-circuit current. The maximum power at the output of a module is determined by equation [12]:

$$P_{max,3} = I_{max,3} \times V_{max,3} \tag{12}$$

For a number of modules N_s in series and N_p in parallel, the maximum power produced is given by (equation [13]):

$$P_{max} = N_p.N_s.I_{max,3}.V_{max,3} \tag{13}$$

2.2.4. Model 4: Jones and Underwood Simulation Model (Fill Factor Power Model)

The following model was developed by Jones and Underwood in 2002 to calculate the maximum power delivered by a photovoltaic module.

$$P_{max,4} = FF.\left(I_{cc}.\frac{G}{G_0}\right).\left(V_{co}.\frac{\ln(k_1 G)}{\ln(k_1 G_0)}.\frac{T_0}{T_c}\right) \tag{14}$$

The coefficient K_1 is calculated by the following formula:

$$k_1 = \frac{K}{I_0} \tag{15}$$

Where I_0 is the inverse saturation current of the diode and K is a constant expressed in $\frac{\frac{A}{W}}{m^2}$ defined by:

$$K = \frac{I_{cc}}{G_0} \tag{16}$$

FF is the fill factor defined by:

$$FF = \frac{P_m}{V_{co} \times I_{cc}} = \frac{V_{mp} \times I_{mp}}{V_{co} \times I_{cc}} \tag{17}$$

P_m is the maximum power of the module under standard test conditions.

2.2.5 Model of Prediction of the Power Produced by the Photovoltaic Generator of Sévagan

In the literature, several authors have carried out comparative studies between the different power models. The reference model is model 1 which was developed experimentally by Lu Lin in 2004. Lu Lin has checked and compared the model of Jones and Underwood and those of Borowy and Salameh. He found that the model of Borowy and Salameh gives a very good accuracy when solar irradiation and temperature are high ($G \geq \frac{900W}{m^2 et T_c} \geq 27°C$). Belhadj Mohammed then compared the three simplified power models to the reference model.

According to his results, the values delivered by the model input/output (Model 2) are in very good agreement with the reference model. In addition, the error is linear as a function of temperature and irradiation, this error does not exceed 0.5 per cent in all cases (Kashif Ishaque, 2011). The input/output model was therefore used to estimate the power produced by the Sévagan photovoltaic generator.

2.3 Model to Describe the Behavior of Photovoltaic Cell

A photovoltaic cell is an optoelectronic component made of semiconductor materials doped differently: type P and type N. It transforms solar light into electricity by photovoltaic effect.

The models used to describe the behavior of a real photovoltaic cell are the one-diode and two-diode models. The one-diode model is easy to implement but less accurate than the two diode model. The iterative method described in the work of Villalva (M. Villalva, 2009) is the best. However, its accuracy deteriorates for low irradiations, particularly in the vicinity of the open circuit voltage (Kashif Ishaque, 2011). The introduction of the second diode in the circuit increases the number of parameters to be determined: the inverse saturation current and the ideality factor of the second diode are added to the existing parameters. The determination of all parameters of the model constitutes the major problem.

For the two-diode model, the Levenberg-Merquardt fitting technique is applied to construct the I-V curve, (Ja Gow, 1996, 1999). According to the work of (S. Chowdhury, 2007), an equivalent Thevenin circuit is used to estimate the model parameters, whereas according to Hovinen (Anssi Hovinen, 1994), the parameters are calculated as a function of the series resistance. However, in all these techniques, (Hovinen, 1994; J. Hyvarinen, 2003), several new additional coefficients are introduced into the equations. In addition, difficulties arise in determining the initial values of the parameters. Another approach to describe the two-diode model is to examine its physical characteristics such as electrons defect coefficient, lifetime of minority carriers, carrier intrinsic density, and other semiconductor parameters. The most important work in this field is the method of decomposing the irradiation of the solar cell (J. Hyvarinen, 2003) (the irradiance decay cell analysis method), the zone of predominance of the diffusion current (Ken-ichi Kurobe, 2005), the two-diode model modified (Kensuke Nishioka, 2003) and the modified three-diode model (Kensuke Nishioka, 2007). Although these models are useful for understanding the physical behavior of a solar cell, information on semiconductor parameters is not always available on the technical data sheets of commercially available photovoltaic cells. Moreover, in most of these works, the following values have been considered for ideality factors: $n_1 = 1$ and $n_2 = 2$. This assumption is widely used but is not always true. In view of these discussions, we can conclude that, although the two-diode model is preferable in terms of precision, its implementation requires more computation compared to the one-diode model, (Hovinen, 1994; J. Hyvarinen, 2003). Moreover, the modeling using the semiconductor approach as described in (Kensuke Nishioka, 2003) is not suitable because of the insufficient information on the technical data sheets. The method proposed by (Kashif Ishake, 2011) has the advantage of simplifying the equation of current: only four parameters have been calculated. The equivalent electrical circuit of the model is shown in Figure 16.2.

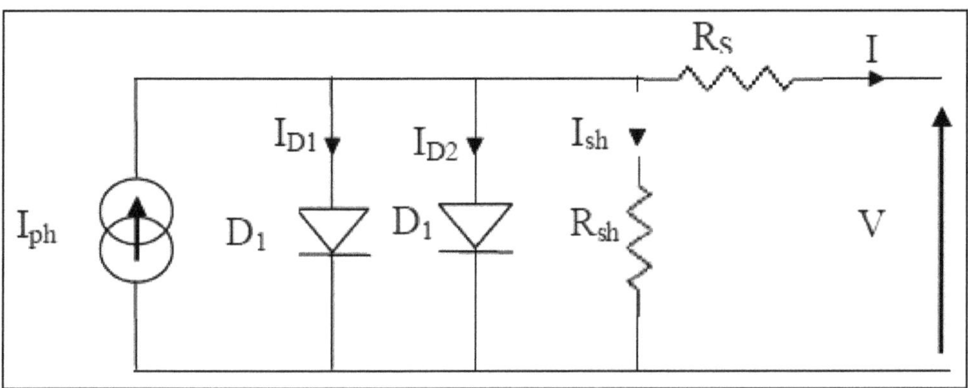

Figure 16.2: Equivalent Electrical Circuit of Two Diodes Model.

The output current is given by Equation 18:

$$I = I_{ph} - I_{01}\left[\exp\left(\frac{(V + R_s I)}{n_1 . V_t}\right) - 1\right] - I_{02}\left[\exp\left(\frac{(V + R_s I)}{n_2 . V_t}\right) - 1\right] - \frac{V + R_s I}{R_{sh}} \tag{18}$$

Where I_{01} and I_{02} are the inverse saturation currents of diode D_1 and diode D_2 respectively. n_1 and n_2 represent the ideality factors of the diodes. The term containing I_{02} in equation 18 compensates for loss by recombination in the space charge zone as described in (Chih-Tang SAH, 1957). For the determination of the parameters, we rely on the model proposed by (Ishaque Kashif, 2001) inspired by that of (Villalva, 2009). The current is given by Equation 22. A modification of Equation 23 gives the inverse saturation currents I_{01} and I_{02} which are taken as equal.

$$I_{01} = I_{02} = I_0 = \frac{I_{cc,n} + K_I \Delta T}{\left(\frac{V_{co,n} + K_V \Delta T}{[(n_1 + n_2)/p].V_t}\right) - 1} \tag{19}$$

Factors n_1 and n_2 represent the components of the diffusion and recombination current, respectively. In accordance with Shockley's theory of distribution, n1 must be equal to unity, (Chih-Tang SAH, 1957). The value of n_2 can vary. A better compatibility is obtained between the proposed model and the experiment if $n_2 \geq 1.2$. Since $(n_1 + n_2)/p = 1$, it follows that $p \geq 2$. This generalization eliminates the ambiguity in the choice of the values of n_1 and n_2. Equation 18 can be simplified in terms of p as:

$$I = I_{ph} - I_O\left[\exp\left(\frac{(V + R_s I)}{V_t}\right) + \exp\left(\frac{(V + R_s I)}{(p-1)V_t}\right) - 2\right] - \frac{V + R_s I}{R_{sh}} \tag{20}$$

Rs and Rsh are calculated simultaneously in a manner similar to the procedure proposed by (M. Villalva, 2009); The maximum power calculated by the model I - V using equation 21, $P_{max,m}$ is equal to the maximum experimental power $P_{max,e}$ supplied by the manufacturer on the datasheet. Thus, $P_{max,m} = P_{max,e} = V_m \; I_m$ gives us the equation 21 for the shunt resistance:

$$R_{sh} = \frac{V_m + I_m R_s}{\left\{I_{ph} - I_O\left[\exp\left(\frac{(V_m + R_s I_m)}{V_t}\right) + \exp\left(\frac{(V_m + R_s I_m)}{(p-1)V_t}\right) - 2\right] - \frac{P_{max,e}}{V_m}\right\}} \tag{21}$$

The series resistor is initialized to zero and the value of the series resistance is set to zero. The shunt resistance initialy is given by equation 24. The iterative method of determining Rs and Rsh is represented by the flowchart of Figure 16.3; The Newton Raphson method will be applied to equation 20. The variable p can take any value greater than 2.2. Despite the high number of parameters, the model proposed by (Kashif Ishaque, 2011). only allows us to calculate four since $I_{01} = I_{02} = I_0$; $n_1 = 1$ and the value of p can be chosen arbitrarily, ie $p \geq 2.2$.

$$I_{ph} = (I_{ph,n} + K_I \Delta T) \times \frac{G}{G_n} \tag{22}$$

$$I_o = \frac{I_{cc,n} + K_I \Delta T}{\exp\left(\frac{V_{co,n} + K_V \Delta T}{n V_t}\right) - 1} \tag{23}$$

$$R_{sh,0} = \frac{V_m}{I_{cc,n} - I_m} - \frac{V_{co,n} - V_m}{I_m} \tag{24}$$

3. Results and Discssions

In this section, experimental data measured on site were compared to simulated power values. The parameters required for the calculation of the power are: the ambient temperature in the vicinity of the cells and the solar irradiation. These data are stored automatically by the ALMEMO control unit.For the experimental power values, voltmeter and an ampermeter were used to take the current and voltage measurements. The photo-voltage and photo-current values were taken every ten (10) minutes. The check was carried out for two typical days: a cloudy day (first day: Saturday 03 December 2016) and clear sky conditions with a few cloudy

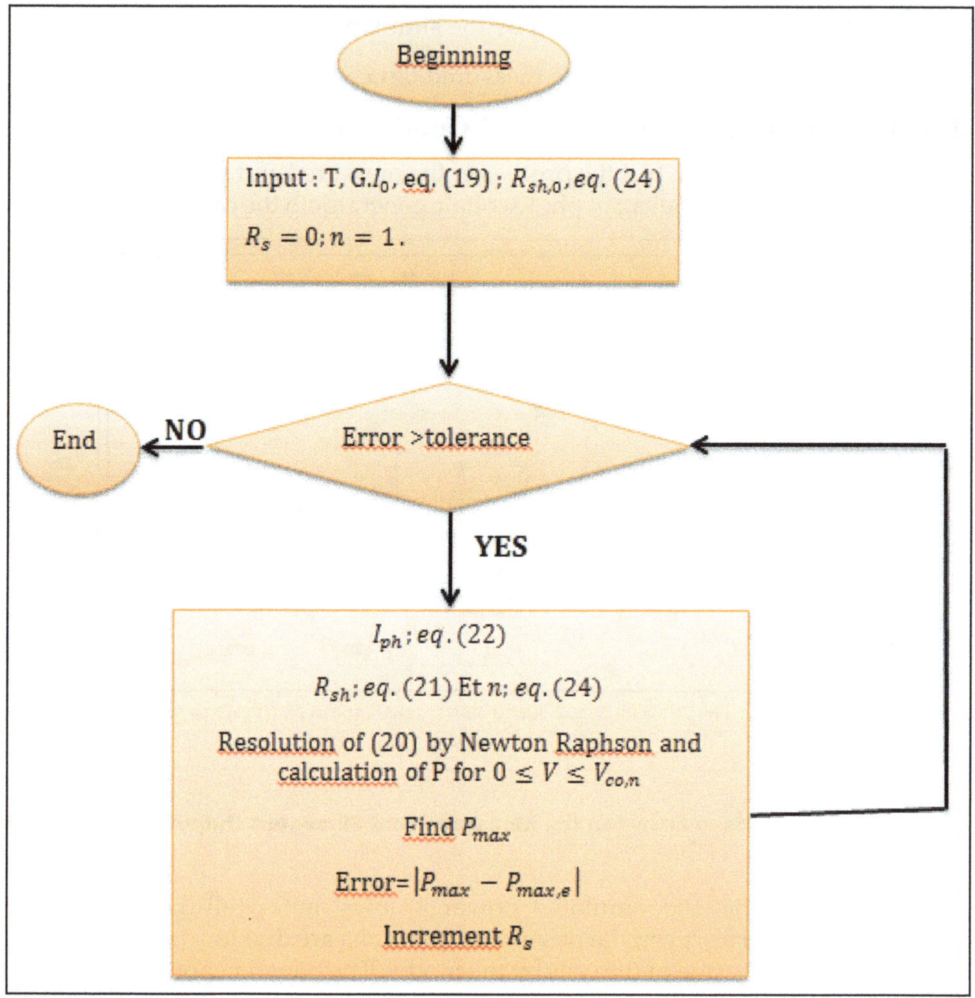

Figure 16.3: Flowchart of Resolution of Equation (20).

passages (day 2: Sunday 04 December 2016). In each case, the input/output model was compared to the experimental data. This allowed us to verify the validity of the model in the full range of weather conditions. At least 60 values for each of the clear sky and cloudy conditions were used to verify the simulation model. The comparison of the results is facilitated by the implementation of a linear correlation coefficient R^2 given by equation (25) (Whei Zhou, 2007):

$$R^2 = 1 - \frac{\Sigma(y_i - \widehat{y_i})^2}{\Sigma(y_i - \overline{y})^2} \tag{25}$$

With:

Y_t represents the measured data of the current of the photovoltaic generator;

$\hat{y}_i$ is the current value predicted by the simulation model;

$\bar{y}$ is the arithmetic mean of the measured data.

3.1. Model Check: Case of Cloud Conditions

We present in Figure 16.4 the profiles of the measured and simulated values of the output power of the Sévagan photovoltaic generator in the case of a cloudy day.

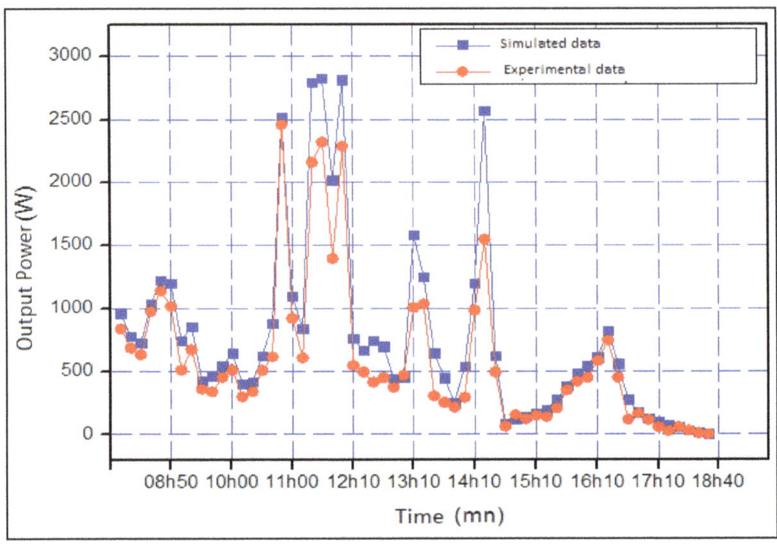

Figure 16.4: Comparison between the Measured and Simulated Output Power Values for a Cloudy Day.

We observe that the simulated power follows quite well the trend of the measured values. However, the observed deviations are due to a poor estimation of the temperature value of the cells by the model. The phenomenon of ambient air convection, thermal inertia and/or poor location of the thermal sensor may cause a delay between the measured ambient temperature used to predict the temperature of the cell and its actual temperature. We can emphasize that the values of the photocurrent and the voltage delivered by the photovoltaic generator are not measured simultaneously. Also, an increase in the error on the measured power, due to the indirect measurement by product I V.

We also present on (Figure 16.5) the correlation between the simulated values and the measured values of the power.

We see clearly that the calculated values vary linearly with the experimental data. Only slight differences were observed. The red line corresponds to the points where the calculated values coincide with the experimental values. The correlation coefficient thus calculated gives us a value of 0.95.

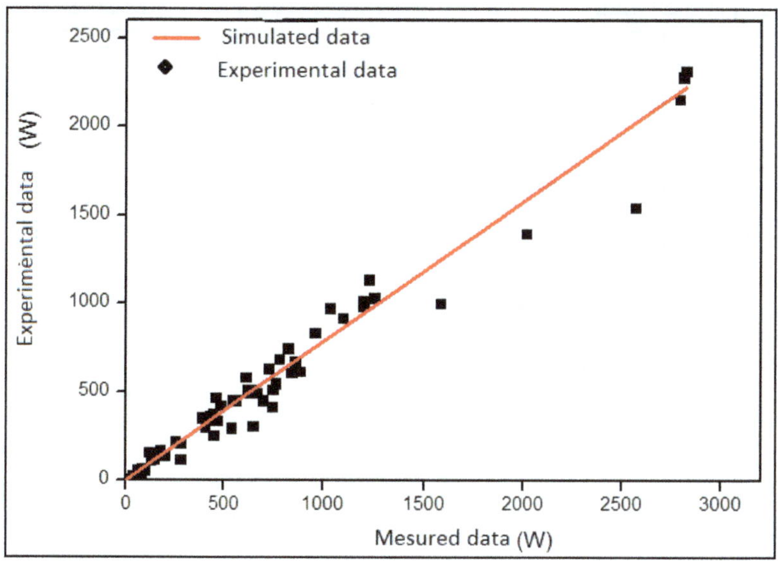

Figure 16.5: Correlation between Measured and Simulated Power Values (Cloud conditions).

By way of comparison with the literature we present on (Figure 16.6) the results obtained by (Whei Zhou, 2007) in the case of the prediction of the performance of a photovoltaic system integrated in the building.

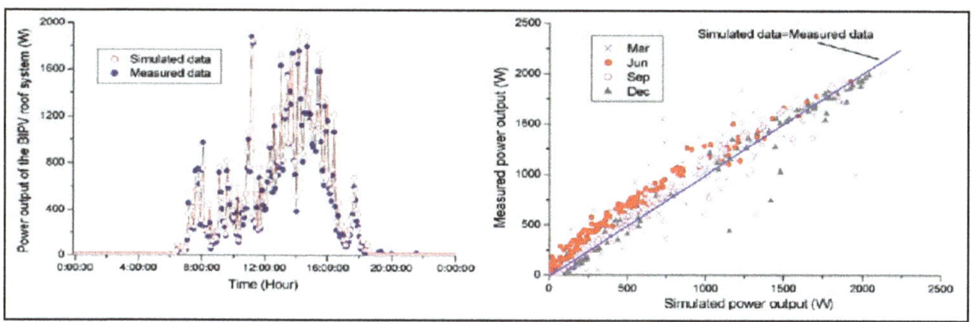

Figure 16.6: Results Obtained by Whei Zhou *et al.* (Cloud Conditions).

In the work of (Whei Zhou, 2007) correlation curves were plotted for four typical cloudy days of March, June, September and December. They obtained a linear correlation coefficient R^2 of 0.96.

3.2. Model Check: Clear Sky Day with only a Few Cloudy

The data from our second measurement day in Sévagan were used to check the power model. We show in Figure 16.7 the profiles of the measured and simulated values of the output power. We find a good agreement of the simulation model with the experimental values. The deviations observed in this case are smaller compared

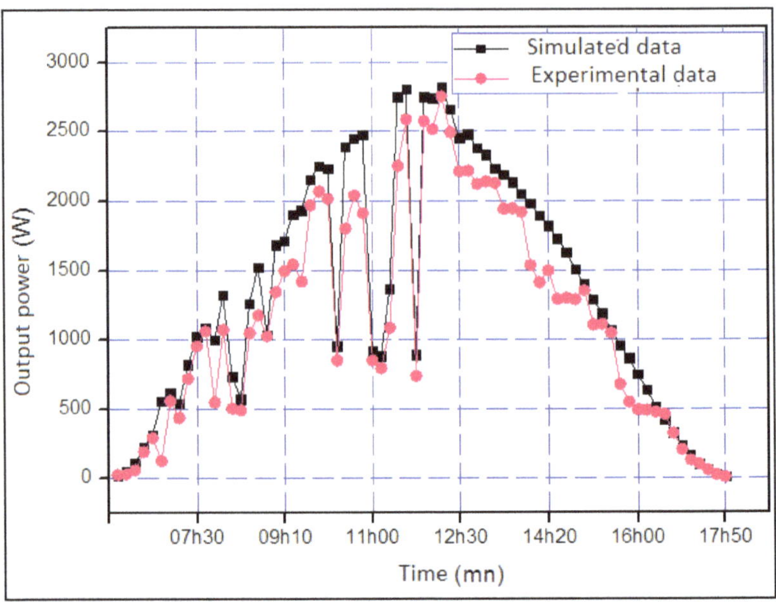

Figure 16.7: Comparison between the Measured and Simulated Power Values (Clear Sky Conditions).

to the cloudy day. Figure 16.8 shows the correlation between the measured values and the theoretical values. The correlation is linear with a correlation coefficient R^2

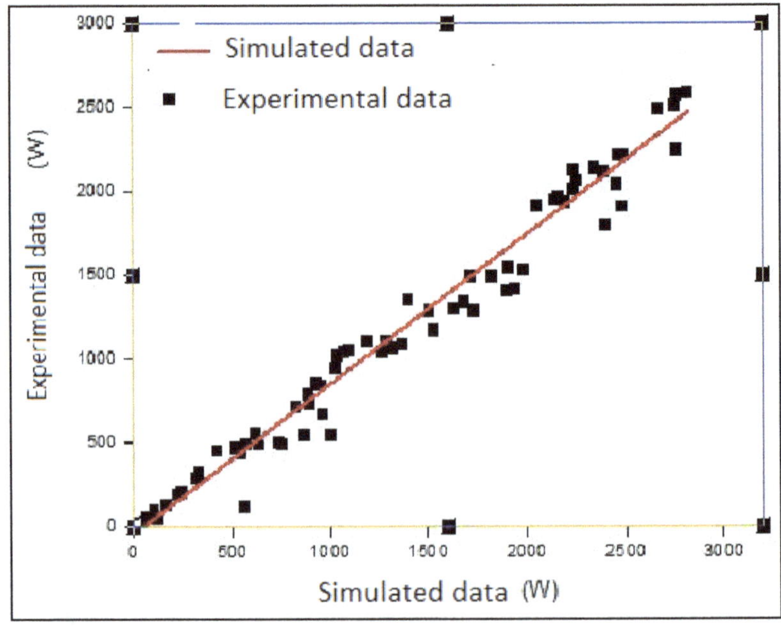

Figure 16.8: Correlation between Measured and Simulated Power Values (Clear Sky Conditions in Togo).

= 0.97. This correlation coefficient is higher than that obtained for the first cloudy day. This difference can be explained by a greater error on the estimated value of the temperature of the modules under cloudy conditions. Indeed, the rapid changes in the solar irradiation caused by the cloudy passages cause a sudden change in the ambient temperature. Given the thermal inertia of the thermometer used to measure the temperature, there may be discrepancies between the recorded value of the temperature and its instantaneous actual value.

This result is in agreement with that obtained in the literature. We show on Figure 16.7 the results obtained by (Whei Zhou, 2007). In case of sunny days, he obtained a correlation coefficient of 0.98.

In both cases, cloudy or sunny weather, the high values of the correlation coefficient demonstrate a good prediction of the performance of the generator by the simulation model.

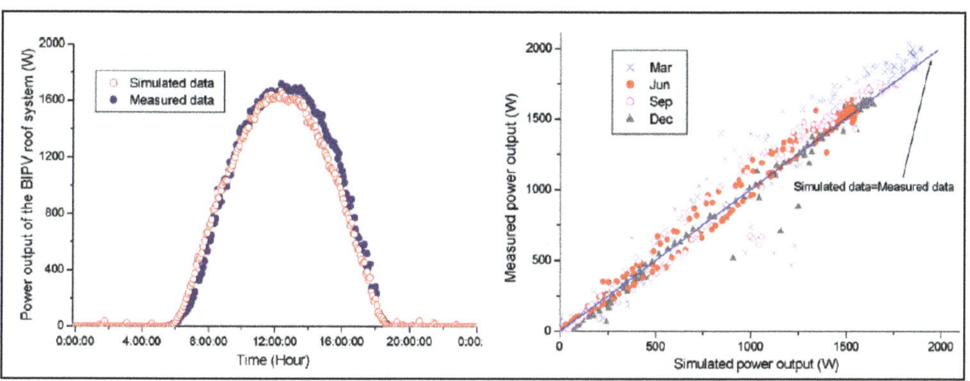

Figure 16.9: Results Obtained by Whei Zhou *et al*. (Sunny Conditions).

3.3. Results for Photocurrent Prediction

The Figures 16.10 show the I-V and P-V curves of the Ecoline LX -260P module which pass exactly through the three experimental points provided by the manufacturer in the technical figure. These curves show the nonlinear relationship between intensity and the voltage of the module on the one hand, and on the other hand between the power and the voltage of the module. They make it possible to understand the operation of a photovoltaic module subjected to a given illumination and a given operating temperature.

The Figures 16.11 and 16.12 show the comparative curves of the simulated and experimental values for different ranges of solar irradiation. These comparisons have been made for the two diode models of the photovoltaic cell. Since solar irradiation and temperature at the module's neighbors are not constant, an average of these magnitudes over the duration of the experiment was considered in the model validation program.

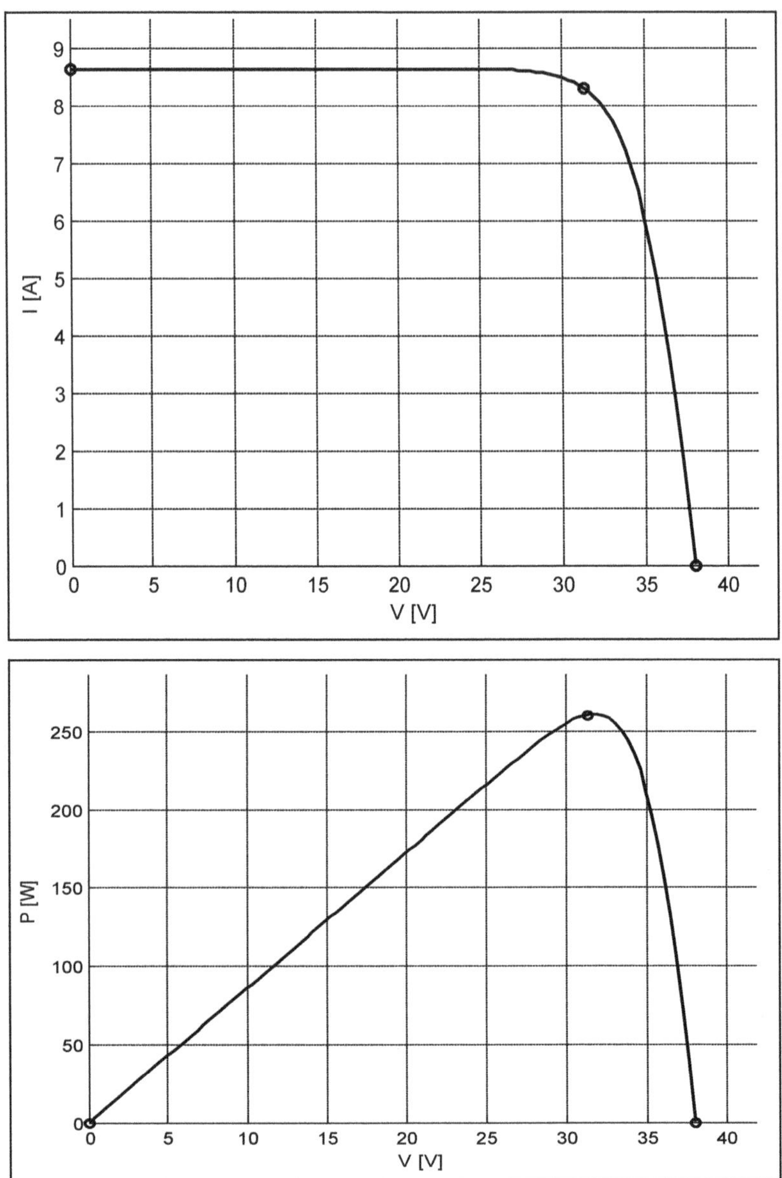

Figure 16.10: Curves I-V and P-V Adjusted to the Three Remarkable Points: Case of the Model with Two Diodes.

In general, good agreement is observed between the experimental curves and the simulation results. However, more or less pronounced deviations are observed. These deviations can be explained by:

- Variation of the luminous flux received by the modulus during an experiment. The light flow received by the module greatly influences the

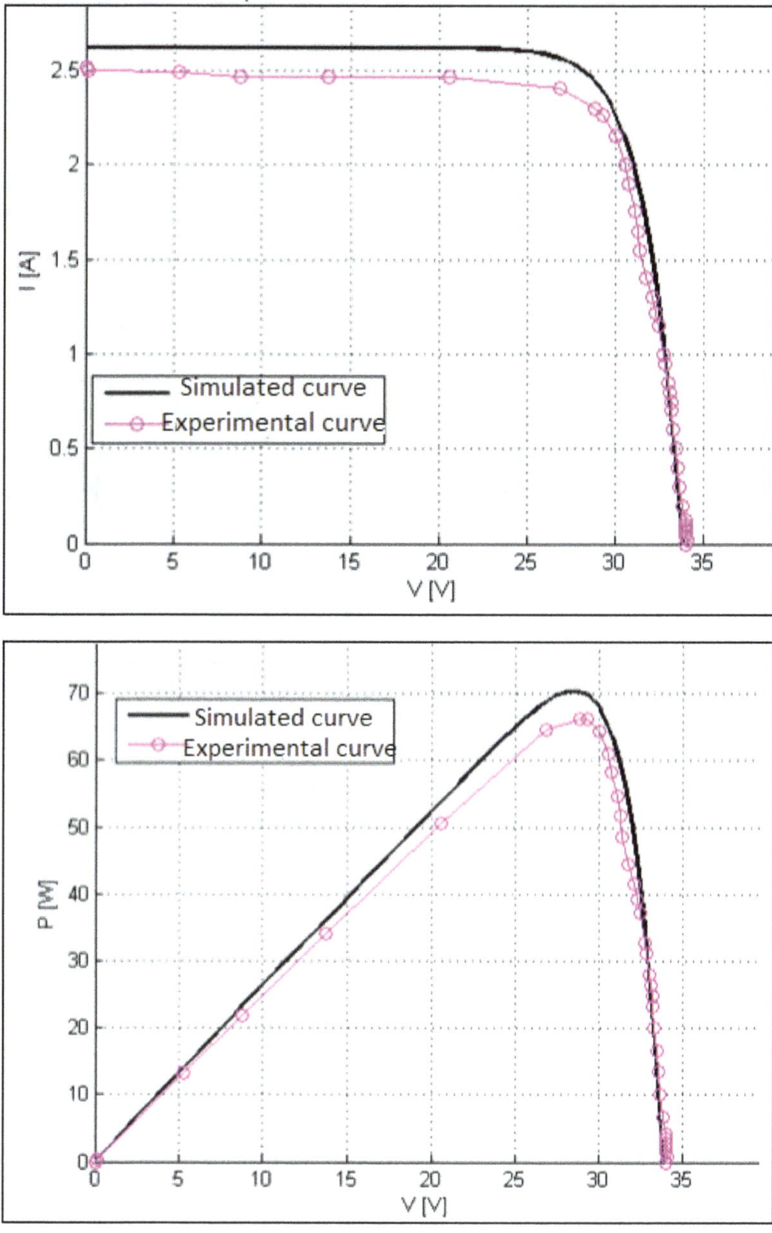

Figure 16.11: Theoretical and Experimental I-V and P-V Curves for G=310W/mi; T=33°C.

photocurrent. This problem can be solved in the laboratory using either a model of characterization of the photovoltaic modules as described in the works of (Bertrand Gelis 2013), or a solar simulator that fixes the illumination per unit area received by the photovoltaic module.

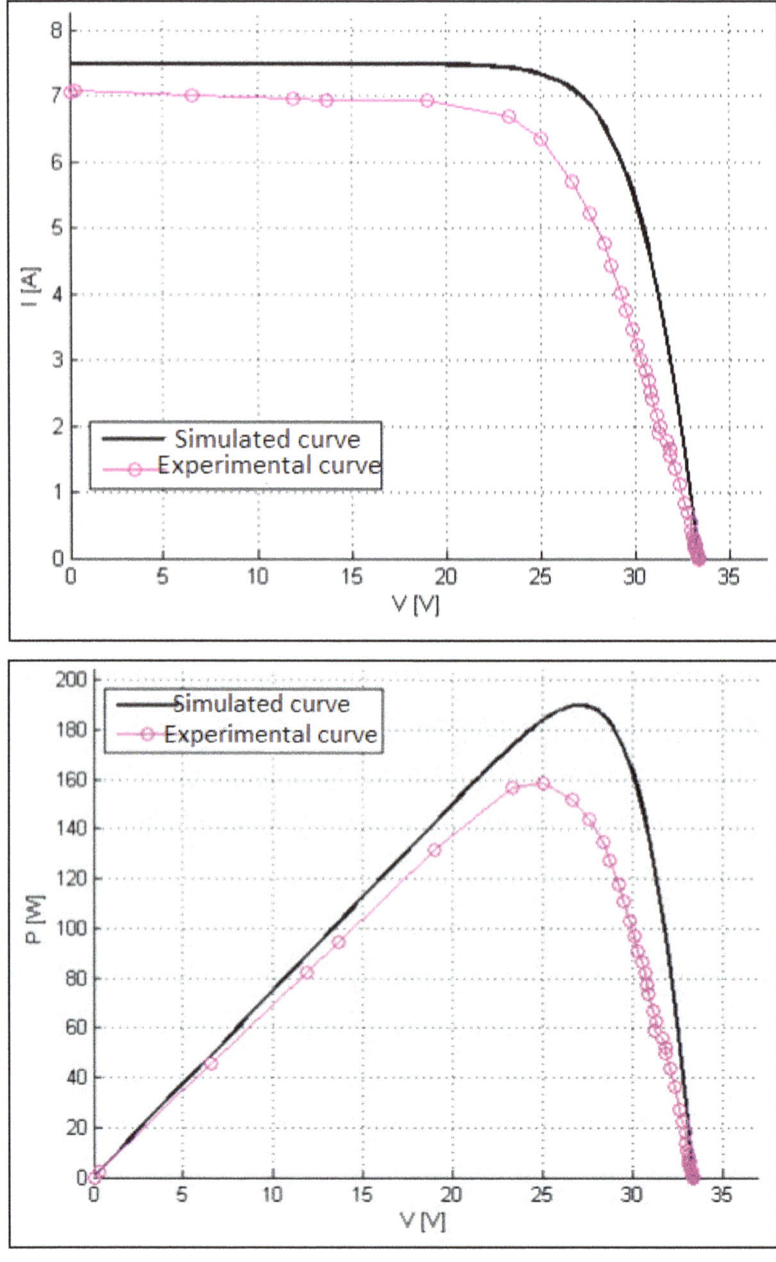

Figure 16.12: Theoretical and Experimental I-V and P-V Curves for G=850W/mi; T=35°C.

- Incorrect estimation of the operating temperature of the cells using the ambient temperature
- The phenomenon of air convection.

- ✰ The thermal inertia of the temperature sensor used may cause a delay between the ambient temperature used to predict the operating temperature of the cells and the actual temperature. Characterization models have a temperature probe to directly measure the temperature of the cells and thus avoid this problem.
- ✰ The performance of the module under real conditions: often the efficiency of the modules under actual conditions of use does not correspond to those provided by the manufacturer.

Presentation of the Irradiation and Temperature Profiles of the Sévagan Site and Validation of the Model

The final objective is to verify whether the model was able to predict the performance of the photovoltaic system. To validate the model, measurements were made over three days (from 03 to 05 December 2016). The total sunshine and the temperature in the vicinity of the modules recorded for these days are shown in Figure 16.13. The solar irradiation increases from the morning to reach a maximum at the solar noon before decreasing again until canceling out at nightfall. The first day of measurement was a very cloudy day with short periods of high solar irradiation. The solar irradiation at Sévagan can exceed $1000\frac{W}{m^2}$; During the three days a maximum of $1150\frac{W}{m^2}$ was reached on the third day at 12h33min.The ambient temperature can drop to about 23.51 C at night, but during the day it depend on the level of sunshine and climatic conditions it can rise between 40 and 45 C.

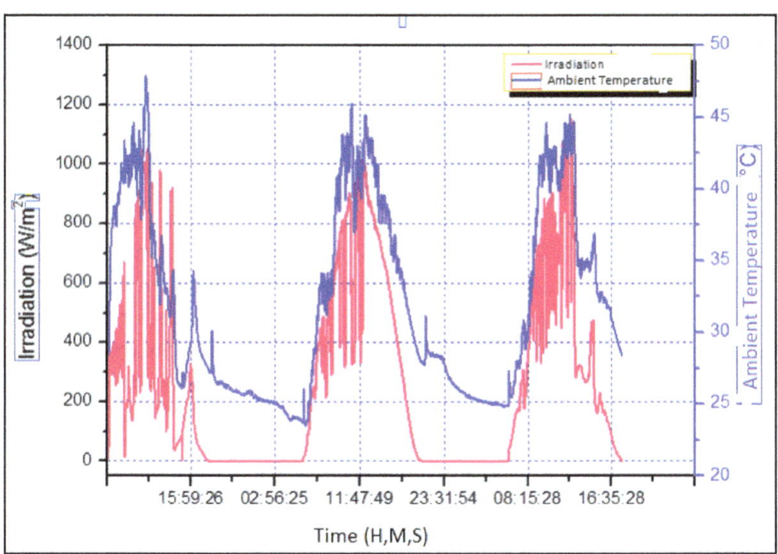

Figure 16.13: Variation of Solar Dadiation and Ambient Temperature in the Vicinity of the Modules for Three Days at the Sũvagan Site.

These types of temperature and solar irradiation profiles have been recorded by several authors, in particular (R. Merahi, 2010 and R. Shenni, 2007). The data were recorded at the Renewable Energy Research Center (CRAER) of the University of Nouakchott, in Mauritania over three days in August (R. Shenni, 2007). The temperature change over the three days showed a maximum temperature of 39 C on the third day. The maximum irradiation on this site during these three days did not exceed $800\frac{W}{m^2}$. The simulation results of the photovoltaic generator were compared with the values measured in (Figure 16.14).

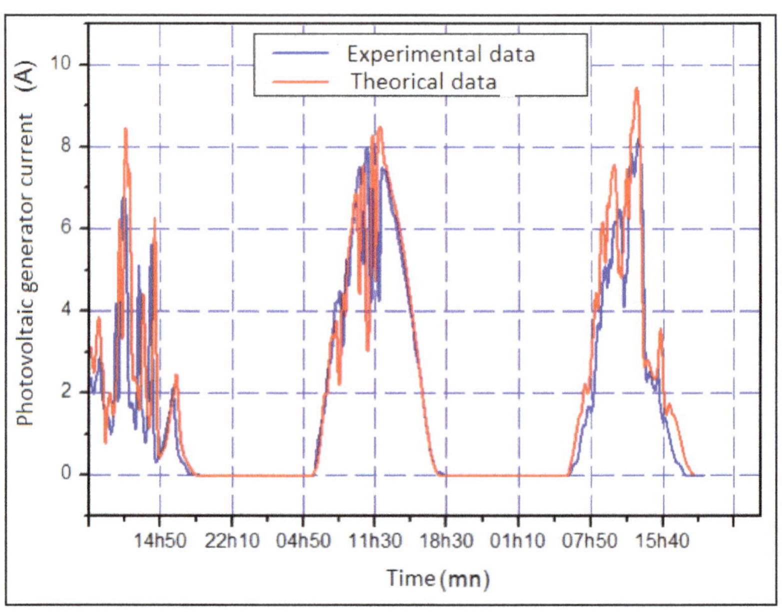

Figure 16.14: Comparison of Measured and Simulated Photocurrent Values.

We find a good agreement for the sunny day with only a few cloudy passages (second day). On the other hand, discrepancies are observed for the cloudy days. These deviations can also be explained by a poor estimation of the temperature value of the cells by the model. The phenomenon of ambient air convection, thermal inertia and/or poor location of the thermal sensor may cause a delay between the measured ambient temperature used to predict the temperature of the cell and its actual temperature.Thus, the rapid changes in the solar irradiation caused by the cloudy passages cause a sudden change in the ambient temperature. Thus, there may be a phase shift between the stored ambient temperature value and its actual instantaneous value, and the values used in the model may not correspond to the actual values.

The correlation coefficients obtained for these three days are respectively: 0.9995; 0.9996 and 0.9963. We note that the sunniest day has the highest correlation coefficient. However, the high values of the correlation coefficient reflect the quality of the model proposed in this study to predict the performance of photovoltaic generators.

4. Conclusions

We present in this paper a list of the models of maximum power of a photovoltaic generator encountered in the literature. Simplified models of maximum power are very useful in practical applications as in the study of photovoltaic systems integrated in the building. On the basis of the work carried out by Lu Lin, 2010 and Belhadj Mohammed, 2010, the input/output model is used to predict the maximum output power of the Sevagan photovoltaic generator in Togo.To verify the validity of the model throughout the range of weather conditions, the verification was done in two steps: on a sunny day and a cloudy day. A good agreement was observed with 95 per cent, 97 per cent and 99 per cent correlation coefficients for cloudy, sunny days respectively and the generator photocurrent simulation. The results demonstrate an acceptable accuracy of the power model under different environmental conditions.

5. Acknowledgements

We present all our acknowledgments to the German Cooperation for the donation and installation of a photovoltaic generator from the ECO LINE LX-260P module at the Sévagan clinic and this, in association with the University of Lomé's Solar Energy Laboratory.

REFERENCES

1. A.D. Jones et C.P. Underwood, 2002. A modelling method for building-integrated photovoltaic power supply. Building Services Engineering Research and Technology, 23(3):167–177.
2. Anssi Hovinen : Fitting of the solar cell iv-curve to the two diode model. Physica Scripta, 1994(T54):175, 1994. 2.1.2.2
3. Ahmed Yahfdhou, Abdel Kader Mahmoud, Issakha Youm, 2016. Evaluation and determination of seven and five parameters of a photovoltaic generator by an iterative method. arXiv preprint arXiv:1601.03257.
4. Belhadj Mohammed, 2008. Modélisation d'un système de captage photovoltaïque autonome. Mémoire de Magistère, Université de Bechar.
5. Bertrand Gélis, Vincent Creuze, Christian Glaize, Franck Lecat et Vincent Thomas : Travaux pratiques de caractérisation de panneaux photovoltaïques. In CETSIS: Colloque sur l'Enseignement des Technologies et des Sciences de l'Information et des Systèmes, numéro 10ème, pages 001–006, 2013. 2.2.2
6. Bogdan S Borowy, Ziyad M Salameh, 1996. Methodology for optimally sizing the combination of a battery bank and PV array in a wind/PV hybrid system. IEEE transactions on energy conversion, 11(2):367–375.
7. Chih-Tang Sah, Robert N Noyce et William Shockley : Carrier generation and recombination in pn junctions and pn junction characteristics. Proceedings of the IRE, 45(9):1228–1243, 1957. 2.1.1, 2.1.2.2, 2.1.2.2

8. Chih-Tang Sah : Fundamentals of solid-state electronics. World Scientic, 1991. 2.1.2.2
9. Fethi Benyarou, 2015. Conception assistée par ordinateur des systèmes photovoltaïques Modélisation, dimensionnement et simulation. Thèse de doctorat.
10. JA Gow et CD Manning : Development of a model for photovoltaic arrays suitable for use in simulation studies of solar energy conversion systems. In Power Electronics and Variable Speed Drives, 1996. Sixth International Conference on (Conf. Publ. No. 429), pages 69–74. IET, 1996. 2.1.2.2
11. JA Gow et CD Manning : Development of a photovoltaic array model for use in power-electronics simulation studies. IEE Proceedings-Electric Power Applications, 146(2):193– Jaakko Hyvarinen et Juha Karila : New analysis method for crystalline silicon cells. In Photovoltaic Energy Conversion, 2003. Proceedings of 3rd World Conference on, volume 2, pages 1521–1524. IEEE, 2003. 2.1.2.2
12. Ken-ichi Kurobe et Hiroyuki Matsunami : New two-diode model for detailed analysis of multicrystalline silicon solar cells. Japanese journal of applied physics, 44(12R):8314, 2005. 2.1.2.2
13. Kensuke Nishioka, Nobuhiro Sakitani, Ken-ichi Kurobe, Yukie Yamamoto, YasuakiIshikawa,YukiharuUraokaetTakashiFuyuki: Analysisofthetemperature characteristicsinpolycrystallinesi solarcellsusingmodied equivalentcircuitmodel. Japanese journal of applied physics, 42(12R):7175, 2003. 2.1.2.2
14. Kensuke Nishioka, Nobuhiro Sakitani, Yukiharu Uraoka et Takashi Fuyuki : Analysis of multicrystalline silicon solar cells by modied 3-diode equivalent circuit model taking leakage current through periphery into consideration. Solar Energy Materials and Solar Cells, 91(13):1222–1227, 2007. 2.1.2.2
15. KR McIntosh, Pietro P Altermatt et Gernot Heiser : Depletion-region recombination insilicon solarcells: whendoesmdr=2. In Proceedings of the 16th European photovoltaic solar energy conference, pages 251–254, 2000. 2.1.2.2, 200, 1999. 2.1.2.2
16. Kashif Ishaque, Zainal Salam et Hamed Taheri, 2011. Simple, fast and accurate two-diode model for photovoltaic modules. Solar Energy Materials and Solar Cells, 95(2):586–594.
17. Lin Lu et HX Yang, 2004. A study on simulations of the power output and practical models for building integrated photovoltaic systems. Journal of solar energy engineering, 126(3):929–935.
18. Lin Lu et HX Yang, 2010. Environmental payback time analysis of a roof-mounted building integrated photovoltaic (bipv) system in hong kong. Applied Energy, 87(12):3625–3631.
19. M Belhadj, T Benouaz, A Cheknane et SMA Bekkouche, 2010. Estimation de la puissance maximale produite par un générateur photovoltaïque. Revue des énergies renouvelables, 13(2):257–264.

20. R Chenni, M Makhlouf, T Kerbache A Bouzid, 2007. A detailed modeling method for photovoltaic cells. Energy, 32(9):1724–1730.
21. R Merahi, R Chenni et M Houbes, 2010.Modélisation et simulation d'un module pv par matlab. Journal of Scientific Research N vol, 1.
22. R Zieba Falama, A Dadjé, N Djongyang et SY Doka, 2016. A new analytical modeling method for photovoltaic solar cells based on derivative power function. Journal of Fundamental and Applied Sciences, 8(2):426–437.
23. S Chowdhury,GA Taylor,SP Chowdhury,AK Saha etYH Song : Modelling, simulation and performance analysis of a pv array in an embedded environment. In Universities Power Engineering Conference, 2007. UPEC 2007. 42nd International, pages 781–785. IEEE, 2007. 2.1.2.2
24. Tanvir Ahmad, Sharmin Sobhan et Md Faysal Nayan, 2016. Comparative analysis between single diode and double diode model of pv cell: Concentrate different parameters effect on its efficiency. Journal of Power and Energy Engineering, 4(03): 31.
25. Wei Zhou, Hongxing Yang, Zhaohong Fang, 2007. A novel model for photovoltaic array performance prediction. Applied energy, 84(12):1187–1198.
26. Weidong Xiao, William G Dunford, Antoine Capel, 2004. A novel modeling method for photovoltaic cells. In Power Electronics Specialists Conference. PESC 04, volume 3, pages 1950–1956.

Chapter 17

Quantitative Analysis of Commercial Solar Cells and Photovoltaic Modules with Electroluminescence

Okan Yılmaz

Alp Osman Kodolbaş, TÜBİTAK,
Marmara Research Center
E-mail: alposman.kodolbas@tubitak.gov.tr

I/V characterization under solar simulator illumination of commercial photovoltaic (PV) modules are used for their classification during production. Such classification is important for both estimation of electrical energy production and improvement of the quality of the modules type approved according to international standards. EL images of the modules are complementary to I/V measurements and are not quantified. In this way, it is not possible to quantify between the EL images of two modules manufactured using different materials (*i.e.* solar cell, tabbing ribbon, flux) or production conditions. On the other hand, EL images of the solar cells are widely used in their development. In this study, we calculated series resistance and open circuit voltage from analysis of EL images of crystalline silicon solar cells based on arguments about local diode voltage and we propose a new numerical method to quantify between the EL images of different modules.

Keywords: *Photovoltaic module, EL imaging, Open circuit voltage mapping, Statistical process control method.*

1. Introduction

Photovoltaic (PV) module yield for years is among the most important factor in determining the cost of solar electricity, besides the capacity of a photovoltaic solar installation, system price and the annual solar irradiance at the installation site. Electroluminescence (EL) imaging is frequently used to identify failure mechanisms of PV modules after particularly in string interconnection and lamination stages of the production [1]. Using I/V characterization under AM1.5G solar simulator illumination photovoltaic modules are then classified in 5W tolerances. EL images of the modules could help to identify the power differences but data mostly stored for future possible assistance. On the other hand, EL imaging in solar cells allows investigation of diverse physical properties and their development. Those physical parameters includes but not limited to series resistance [2], external quantum efficiency [3-5].

Local series resistance R_S (x,y) is defined in different ways. In the current R_S imaging methods R_S is defined as the local voltage drop between the bias voltage (V) applied to the busbars and the local diode voltage V_d (x,y), divided by the local diode current density J_d (x,y):

$$R_s(x,y) = \frac{V - V_d(x,y)}{J_d(x,y)} \tag{1}$$

Here the dark diode current is dened as positive and the photocurrent as negative. Also, R_S has the unit of W cm^2. This denition was used right describes the global series resistances of cells of different size, thereby ensuring that the series resistance is homogeneous and independent of the area A of the cell. On the other hand for inhomogeneous solar cells which includes regions of higher current density or contact resistance the model is applied for each pixel of a solar cell image [6]. Under this condition local EL intensity is described as

$$I_{EL,i} = C_i \exp\left(\frac{V_i}{V_t}\right) \tag{2}$$

Here, C_i is the local proportionality factor and V_t is thermal voltage kT/e [3]. Applying the Fuyuki approximation and following iteration scheme local diode voltage is calculated [3] using

$$V_{i,1}^{(2)} = V_i - \frac{R_{s,i}^{(1)} J_{0,i}^{(1)} \exp\left(V_{i,1}^{(1)}\right)}{V_t} \tag{3}$$

This paper is intended in two parts; in the first part, arguments about the series resistance and local diode voltage calculations have been applied to commercial solar cell. In the second part, based on EL image intensity measurements and statistical methods we have proposed a new numerical method to quantify between the EL images of different modules.

2. Materials and Methods

A TE cooled Si-CCD camera and IR filters was used to capture EL image. In total 34 modules are examined in the measurements. Multi crystalline modules supplied from a manufacturer in Turkey. Their number was 33. Remaining mono crystalline silicon module is manufactured in TÜB TAK Photovoltaic Technology Center. In mono crystalline and multi crystalline modules typical 156x156 mm^2 sized industrial cells have been used. Efficiency of cells, U_{OC}, I_{SC} and P_{MPP} have been, 18.9 per cent,0.629 V, 9.080 A and 4.484 W for mono crystalline cells and 17 per cent, 0.627 V, 8.243 A and 4.04 W for multi crystalline cells. During the EL measurements temperature was kept constant at 23±1°C and RH at 45 per cent ±5. STC performance parameters of all the multi crystalline modules were labeled the same by the manufacturer. For the monocrystalline silicon module results of the STC measurements are labeled. EL images of all the cells and modules were taken at indicated V_{OC} values and at two additional bias voltages for the calculations [6]. Local diode voltages were calculated using the method suggested in [3]. Camera software is used for the analysis of EL images. Statistical Process Control Method (SPCM) is used to for the first time for the classification of the modules through EL images [7].

3. Results

Figure 17.1 shows EL images of mono and multi crystalline silicon cells and Figure 17.2 shows EL images of mono and one of the multi crystalline silicon modules manufactured from cells in Figure 17.1. As mentioned, EL imaging is powerful tool to distinguish in solar cells and module production. In this way, grain boundaries, crystallographic defects, cracks, broken fingers, screen printing errors in solar cells and poor stringing, misalignments in modules are distinguishable.

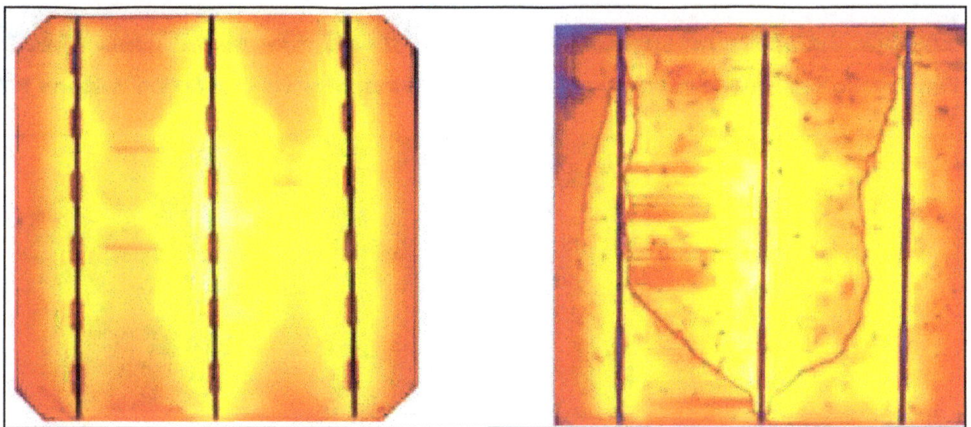

Figure 17.1: Electroluminescence Images of a) Mono, b) Multi Crystalline Silicone PV Cells.

Figures 17.3 and 17.4 respectively shows EL images of the mono and multi crystalline silicon cells and local diode voltage (Local V_{OC} image), R_S, and J_d of the photovoltaic cells for 1st and 20th iterations calculated using Eqn. [3]. By increasing iteration number more reliable results to use in quantitative analysis of EL images were obtained.

Figure 17.5 shows EL images one of the multi crystalline silicon modules and local diode voltage (Local V_{OC} image) of the photovoltaic modules calculated using Eqn. [3]. Calculated average V_{OC} values in Figure 17.5b. is about 2.4V less than the V_{OC} value on the label of the module.

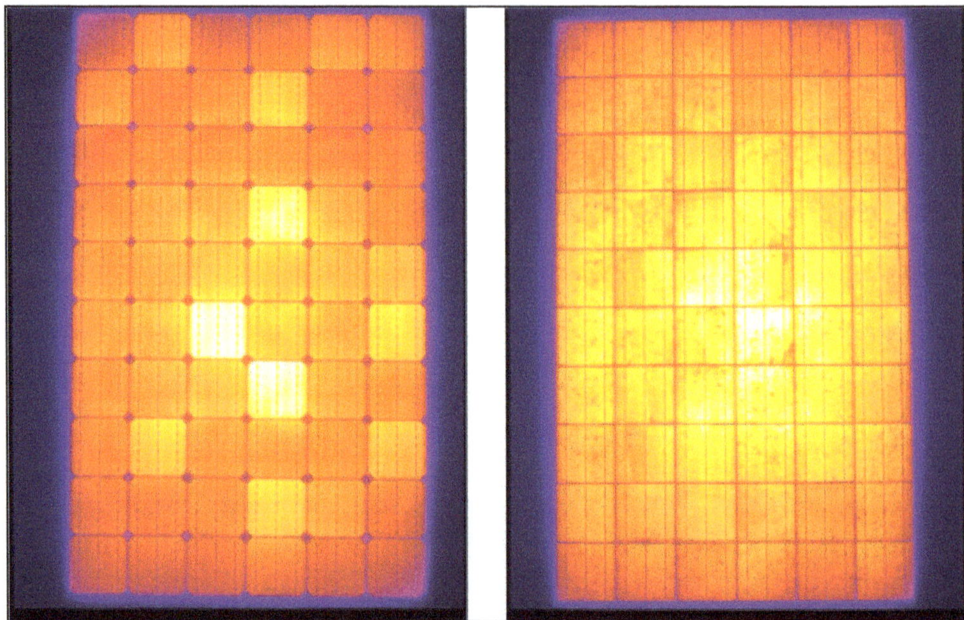

Figure 17.2: Electroluminescence Images of a) Mono, b) Multi Crystalline Silicone PV Modules.

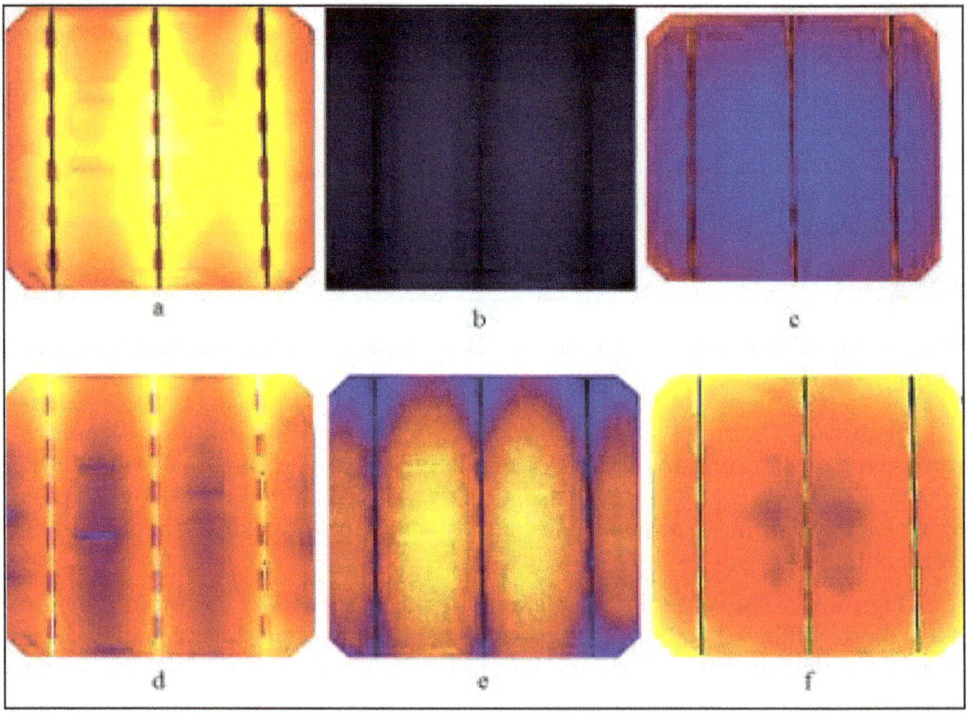

a

b

c

d

e

f

Figure 17.3: Electroluminescence Images of Mono Crystalline Cell after 1th Iteration a) V_{oc}, b) R_s, c) J_0 Mono, after 20th Iteration d) V_{oc}, e) R_s, f) J_0.

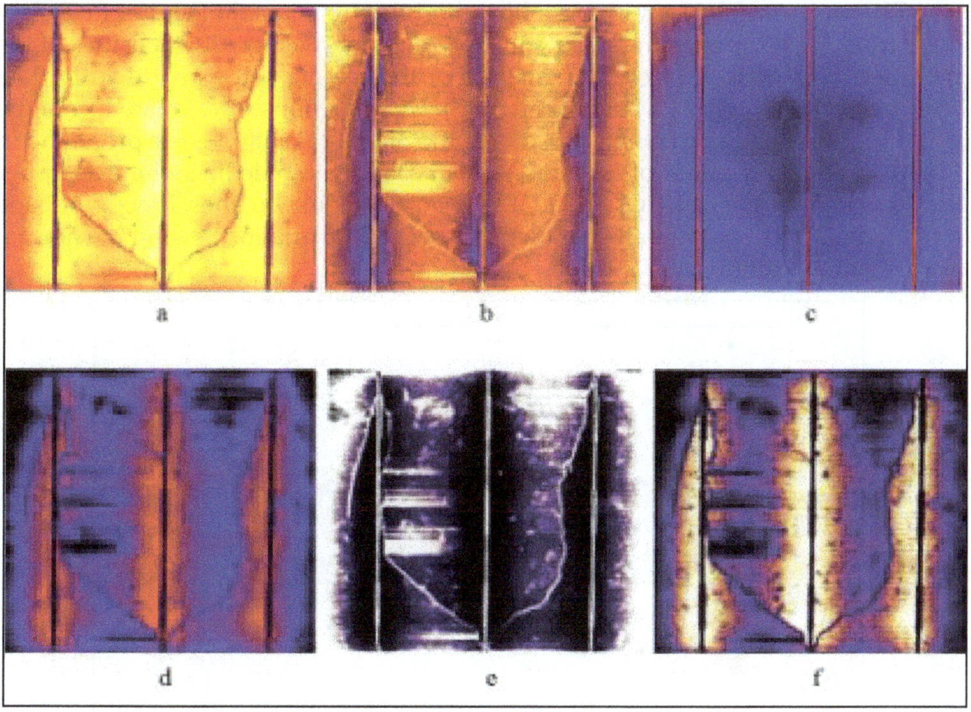

Figure 17.4: Electroluminescence Images of Multi Crystalline Cell after 1th Iteration a) V_{oc}, b) R_s, c) J_0, after 20th Iteration d) V_{oc}, e) R_s, f) J_0.

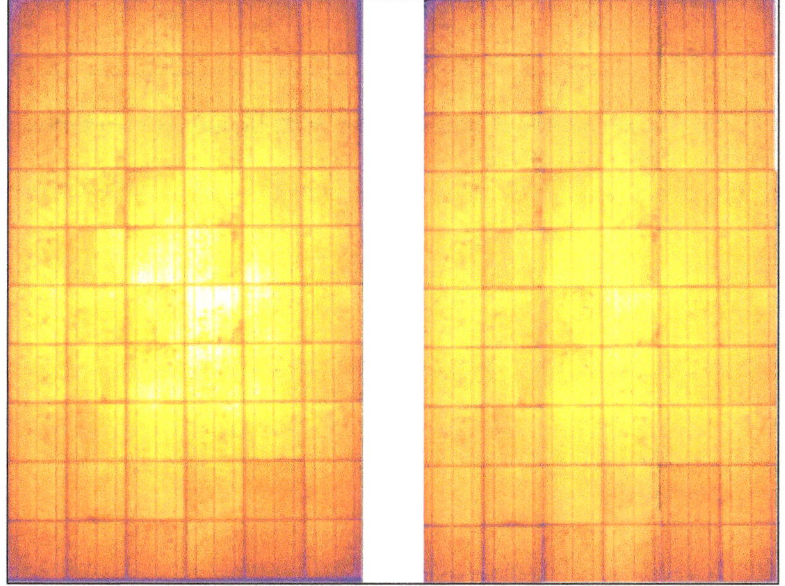

Figure 17.5: Electroluminescence Images of a) One of the Multi Crystalline Module, b) Local Diode Voltages (Local Voc) Images Calculated Using Eqn. [3].

EL images of the 33 multi crystalline modules from a manufacturer in Turkey were used for development of classification method. Average EL intensities of the modules were deduced from the analysis of the camera software. Realized data were analyzed using SPCM described in [7]. EL Subgroups were created from the average of three measurements. Results of the SPCM calculations are presented in Figure 17.6. Control limit of the process is marked with blue line where as upper and lower control limits (UCL and LCL) are marked with red lines. Process outside UCL and LCL is very unlikely and there is a source of variation beyond the normal chance in production.

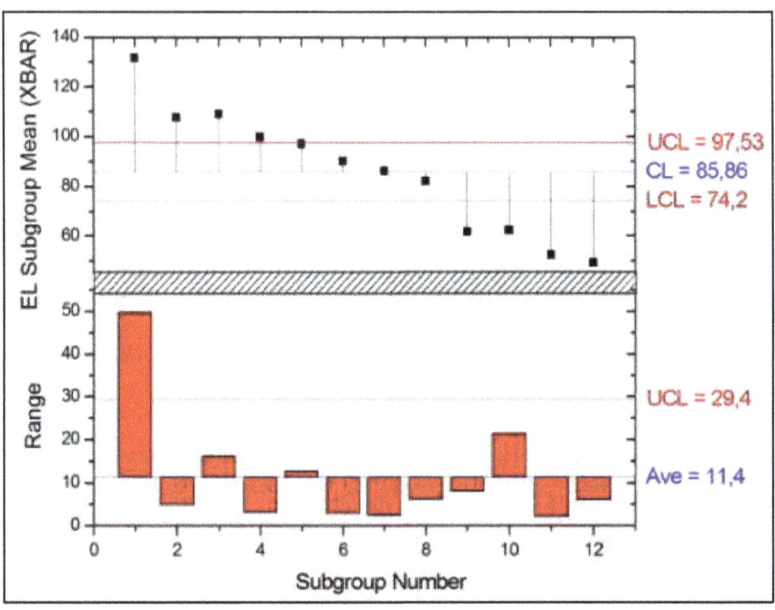

Figure 17.6: Results of the SPCM Calculations a) EL Subgroup Distribution, b) Range of Distribution within the Group.

4. Discussion

EL imaging allows detailed visualization of defects from solar cells and module production. EL image of the monocrystalline module is as expected much more homogeneous than the multi crystalline one leading to 16.8 per cent efficiency in the first one and 15 per cent efficiency for the later. Arguments used to obtain series resistance and open circuit voltage analysis for solar cell are applied for the first time to PV modules. 2.4V difference between the calculated average V_{OC} value and marked one on labeled by the manufacturer suggests that there are resistive loses in the multi crystalline module.

Despite the fact that STC performances of all the modules were marked as same in the labels SPCM calculations using the EL images suggest that there are wide scattering of data beyond UCL and LCL. In other words, if all the modules were all labeled as "A" from I/V measurements, quantitative EL measurements suggest that there are better and also worst modules as marked.

5. Conclusions

It has been shown that calculated local diode voltage (Local V_{OC} image), R_S, and J_d of the photovoltaic cells consistent with literature. Arguments used to calculate series resistance and open circuit voltage from analysis of EL images of solar cells based on independent diode models has been successfully applied to crystalline silicon modules. In this way, defects originating from solar cells and module production are clearly identified and quantified. Statistical analysis of EL is show to be valuable tool for classification and also for process improvement.

6. Acknowledgements

The work is partially supported by project entitled "Fotovoltaik Temelli Güneþ Enerjisi Santral Teknolojilerinin Geliþtirilmesi-M LGES" under the coordination number 113G050. Authors wish to thank A. Seçgin, Y. Vural and T. A. Tumay for their help in production of the monocrystalline module used in this work.

REFERENCES

1. J. Coello. "Introducing electroluminescence technique in the quality control of large pv plants". Proceedings of 26th EUPVSEC (2011) 3469.
2. U. Rau, "Reciprocity relation between photovoltaic quantum efciency and electroluminescent emission of solar cells," Phys. Rev. B, vol. 76, no. 8, p 085303, 2007.
3. O. Breitenstein, A. Khanna, Y. Augarten, J. Bauer, J.-M. Wagner, and K. Iwig, Quantitative evaluation of electroluminescence images of solar cells, Phys. Status Solidi RRL 4, No. 1–2, 7– 9 (2010).
4. D. Hinken, K. Ramspeck, K. Bothe, B. Fischer, and R. Brendel, "Series resistance imaging of solar cells by voltage dependent electroluminescence," Appl. Phys. Lett., vol. 91, no. 18, art. no. 182104, 2007.
5. T. Trupke, E. Pink, R. A. Bardos, and M. D. Abbott, "Spatially resolved series resistance of silicon solar cells obtained from luminescence imaging," Appl. Phys. Lett., vol. 90, no. 9, p. 093506, 2007.
6. F. Frühauf, Y. Sayad, O. Breitenstein, Description of the local series resistance of real solar cells by separate horizontal and vertical components, Solar Energy Materials and Solar Cells 154 (2016) 23–34.
7. Yashchin, E, Statistical Control Schemes - Methods, Applications and Generalizations, International Statistical Review, 61(1) (1993) 41-66.

Chapter 18

Trends in Solar Power Generation and Energy Harvesting in Zimbabwe

Everson Bhunu

Ministry of Higher and Tertiary Education, Science and Technology Development, Zimbabwe
E-mail: bunuz2013@gmail.com

Of all the energy sources, solar power has the most potential to supply all our energy needs for the foreseeable future. This is so because it is infinitely available as it comes from the sun. To date, the declining cost of solar power is competing with that of fossil fuel. In some countries the cost of solar power is cheaper than that of oil. Solar cells are increasingly being more efficient, their price is dropping so fast and the growth of manufacturing is rising so fast. Solar power is now approaching grid parity in some nations across the world. Grid parity occurs when an alternative energy source can generate power at a levelized cost of electricity (LCOE) that is less than or equal to the price of purchasing power from the electricity grid.

Zimbabwe suffers perennial power shortages. According to the Zimbabwe Electricity Supply Authority (ZESA), Zimbabwe needs between 1900 and 2200 megawatts daily but the power utility only manage to generate between 900 and 1200. It imports about 35 per cent of its electricity from Mozambique, South Africa and Democratic Republic of Congo, which still falls short of demand resulting in regular power cuts through load shedding.

Investments in Solar Energy have a greater potential in averting the energy deficits faced by the Zimbabwe. Zimbabwe has one of the best conditions for Solar Photovoltaic (PV) worldwide (solar radiation averaging 20MJ per square metre), with an average of 300 days or 3000 hours of sunshine per year. In September 2015, the Ministry of Energy and Power Development launched a solar energy campaign with hopes to cut residential power usage by 40 per cent through increased use of Solar Energy by the year 2020. Over the years, the government of Zimbabwe, donors, Non-Governmental Organisations and the private sector have supported a number of solar projects to promote thermal and electric applications. Some of the projects include, Solar PV power plants in Gwanda (100MW), Plumtree (100 MW), Munyati (100 MW), Zvishavane (2 MW) and Marondera (150M).

As the Zimbabwean economy continues on the industrialisation and growth phase, the energy demand will continue to rise. There is a need for massive sustainable power investments to meet the current and the anticipated future national power requirements.

This paper seeks to discuss the several applications of solar technologies and the challenges being faced by Zimbabwe in acquiring these technologies. It outlines the solar energy utilisation plans for Zimbabwe. In addition the paper addresses the costs of installation, maintenance, and operation of solar systems and reviews the economic policies that promote the installation of solar energy systems.

Keywords: *Photovoltaic, Economic policy, Load shedding, Grid parity.*

1. Introduction

Zimbabwe is a landlocked Southern African country which shares its borders with Zambia to the northwest, Mozambique to the east, South Africa in the south and Botswana in the southwest, all members of the Southern African Development Community (SADC). Its population is 12.6 million people (2012 National Census). 40 per cent of the population have access to electricity (Zimbabwe Energy Regulatory Authority (ZERA) 2016), with 83 per cent of urban households being connected to the national electricity grid compared to 13 per cent in rural areas. Rural communities

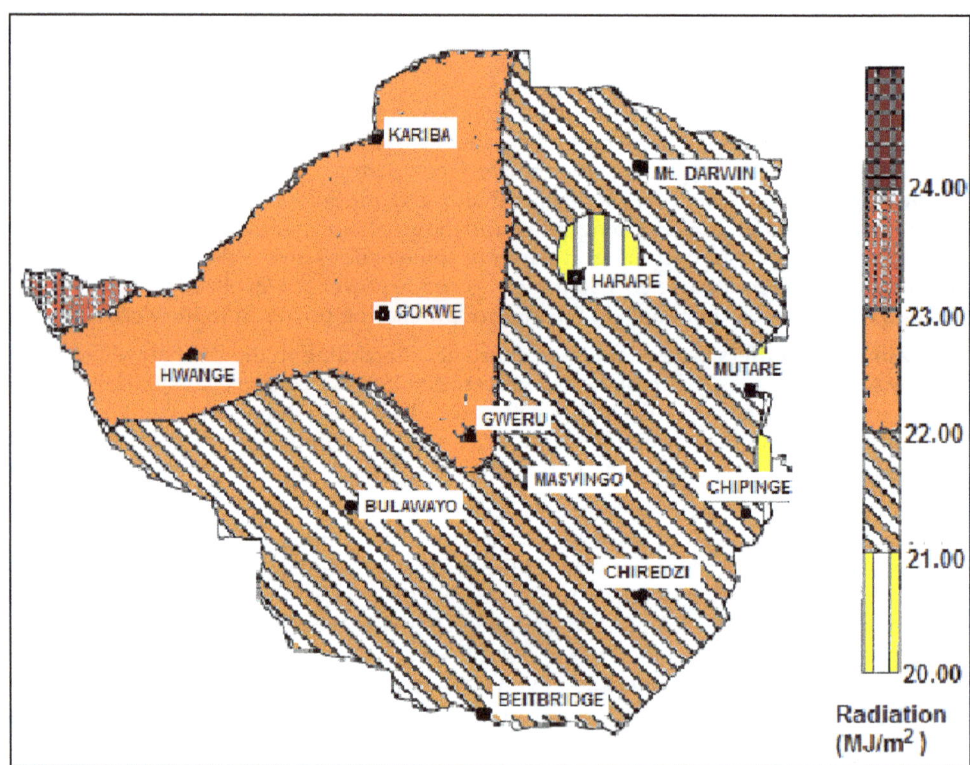

Figure 18.1: Zimbabwe's Annual Mean Radiation (MJ/m²/Day) (*Source: ZERA*).

rely on low quality energy sources such as wood, paraffin/kerosene and candles for lighting and cooking. This lack of access to electricity especially for rural communities, is a barrier towards socioeconomic development. The promotion of the use of alternative sources of energy such as solar, biogas, mini-hydro and biomass help in addressing the above-mentioned challenges.

According to the meteorological institute in Zimbabwe, the country has an annual average of 8.3 sun hours per day; solar radiation averaging 20MJ per square metre, with an average of 300 days or 3000 hours of sunshine per year. These conditions are ideal for the adoption of solar as an alternative source of energy. However, the enormous potential for solar PV and solar water heaters is not being sufficiently exploited.

The demand of solar energy systems in Zimbabwe is very high especially in the remote rural areas where there is no grid electricity. However, the cost of installation of these systems is currently prohibitive. The government is advocating for local production of solar systems in order to reduce the cost. On the other hand, the majority of the population lack knowledge of solar systems that suit their needs. Lack of technical knowhow of these systems results in the procurement of solar systems of low efficiency and reliability. Hence a need for capacity development in this area.

1.1 Solar PV Background

It all began in 1839 when a French physicist, Alexandre Edmond Becquerel discovered the photovoltaic effect (a process in which light is converted into electricity) by coincidence when he was experimenting with metal electrodes and electrolyte. The first selenium solar cell was constructed in 1877. The founding principles of the cell lie on the discovery of the photovoltaic effect in selenium by Willoughby Smith in 1873 complimented by William G. Adams's discovery of the same effect by illumination of the junction between selenium and platinum in 1876. The first photovoltaic cell was built in 1883 by Charles Fritts.

Albert Einstein was awarded a Nobel Prize in 1921 for his theoretical explanation of the photovoltaic effect that was proven in practise by Robert Millikan's experiment in 1916. The method for mono-crystalline silicon production discovered in 1918 enabled mono-crystalline solar cell production. Photovoltaic effect in cadmium-selenide was later observed in 1932. Cadmium Selenide is a very important material for the production of solar cells.

In the period 1950 to 1969 a lot of developments in solar technology occurred. Research results in this era, culminated in the efficiency of the solar cell being increased from 4.5 per cent (1954 Bell's Laboratories results) to around 14 per cent (Hoffman Electronics – USA). The first sun-powered automobile was demonstrated in Chicago, Illinois on August 31st 1955. In the same year, a commercial photovoltaic product with 2 per cent efficiency and 14 mW peak power selling at US$ 25 per cell was produced. The product had an energy cost of US$ 1,785 per watt. The first radiation proof silicon solar cell was produced in 1958. This was designed for Space Technology and during the same year the first solar powered satellite was launched. The period 1960 to 1969 saw the launch of many solar powered satellites by America, Russia and Japan, included is the first commercial telecommunications satellite, Teslar, developed by Bell Industries of America.

The cost of the solar cell was significantly reduced by the works of Dr Elliot Berman and the Exxon Corporation who came up with low cost designs of the solar cell, bringing the price down from US$100 a watt to US$20 a watt in the early 1970s. Progression in Technology resulted in the increased application of solar cells during this era. The solar cells so developed could power navigation warning lights and horns on several offshore gas and oil rigs, railroad crossings, domestic applications and lighthouses. After the 1970 oil crisis, oil companies began to invest in solar research.

The 1980s saw continued improvement on the efficiency of the solar cells and the cost per watt. Water could be drawn from underground using solar power in deserts and solar energy showed up in remote places to power homes. On the other hand, solar cell applications in calculators, radios, lanterns and other battery charging applications became popular. In 1983 the worldwide photovoltaic production exceeded 21.3 megawatts while sales exceeded US$250 million. Continued research in solar photovoltaic technology culminated in the production of reflective solar concentrators that were first used with solar cells in 1989.

Today solar cells have a variety of applications that varies from merely powering a hand calculator or cell phone to powering a plane or running a manufacturing plant. The cost has dropped to an average of $0.80 per watt. The continued advancement in technology today has enabled the development of screen printed solar cells and solar shingles that can be installed on rooftops. The solar technology market is fast growing and many solar panel manufacturers have sprouted around the globe.

2. Key Energy Players in Zimbabwe

2.1 The Ministry of Energy and Power Development

The ministry of Energy and Power Development has the overall responsibility for energy issues in Zimbabwe. The mandate of this ministry includes policy formulation, performance monitoring and regulation of the energy sector as well as research, development and promotion of new and renewable sources of energy. The Ministry also supervises and oversees the performance of state-owned enterprises which include the Zimbabwe Electricity Supply Authority (ZESA), the National Oil Infrastructure Company (NOIC), Petrotrade, and the Rural Electrication Agency (REA).In addition, the ministry regulates Independent Power Producers (IPPs).

2.2 The Zimbabwe Energy Regulatory Authority (ZERA)

ZERA was created for policy monitoring and enforcement. It regulates the production, procurement, importation, transmission, distribution, transportation and exportation of energy derived from any energy source. ZERA has authority for decision making on clearly dened functions that are critical for ensuring operational, nancial and investment eciency in the energy sector. The decision making role of ZERA is concerned with the development, monitoring and enforcement of product and service standards; energy prices; dispute resolution; and the issuing, enforcement, renewal, amendment or cancellation of licences.

ZERA's key objectives include ensuring the security of energy supply, encouraging energy eciency at utility and consumer levels and encouraging use of renewable energy and environmental protection, among others. The Authority is also registering all renewable energy and energy eciency providers operating in Zimbabwe with a view of developing a database and providing recommended suppliers. A number of Independent Power Producers (IPP) have been Licensed by ZERA to date. Eleven of these IPPs are in solar energy for the national grid.

3. Policies

The adoption of Solar Energy is enabled by several policy documents in Zimbabwe. These include the Rural Electrification Act of 2002 which was crafted to spearhead rapid and equitable electrification of rural areas in Zimbabwe. The Energy Regulatory Authority Act of 2011 which established ZERA provides requirements, processes and regulations related to energy resource licensing. The Zimbabwe Agenda for Sustainable Socio-Economic Transformation (ZIMASSET 2013 -2018) also promotes increased uptake of renewable energy. Energy comes under the Infrastructure and Utilities Cluster.

3.1 Zimbabwe National Energy Policy

The National Energy Policy of 2012 promotes the increased uptake of renewable energy resources in the country among others and of course these include solar energy.The policy was launched with the following objectives:

- ☆ Increase access to affordable energy services to all sectors of the economy; through optimal use of available energy resources and diversification of supply options;
- ☆ Stimulate sustainable economic growth by promoting competition, efficiency and investment in the sector;
- ☆ Improve institutional framework and governance in the energy sector to enhance efficiency and energy services delivery;
- ☆ Promote research and development in the energy sector;
- ☆ Develop the use of other renewable sources of energy to complement conventional sources of energy

3.2 The Renewable Energy Feed-in Tari

Developed by the Zimbabwe Energy Regulatory Authority, the renewable energy feed-in tariff scheme is yet to be implemented. It was developed for renewable energy technologies that include solar photovoltaics, mini hydro, biomass, bagasse and biogas for energy projects up to a capacity of 10MW. The renewable energy feed-in tari is a policy instrument that mandates power utilities operating the national grid to purchase electricity from renewable energy sources at a pre-determined price so as to stimulate investment in the renewable energy sector.

3.3 The Zimbabwe Agenda for Sustainable Socio-Economic Transformation

In pursuit of a new trajectory of accelerated economic growth and wealth creation, the Government of Zimbabwe formulated a new plan known as the Zimbabwe Agenda for Sustainable Socio-Economic Transformation (ZIMASSET). ZIMASSET was crafted to achieve sustainable development and social equity anchored on indigenization, empowerment and employment creation. This Results Based Agenda recognises the need for the nation to invest in renewable energy that include biogas, solar, wind, hydropower and biomass.

The Ministry of Energy's tasks in relation to renewable energy as stated in the ZIMASSET are to;

- ✰ Come up with grid code and initiate the programme;
- ✰ Carry out a study on solar water heaters;
- ✰ Develop Policy and regulatory measures on solar water heaters;
- ✰ Initiate a program on solar water heaters.

3.4 Solar PV Integration Code

ZERA developed a solar PV integration code. The code establishes the basic rules, procedures, requirements and standards that govern the operation, maintenance and development of solar PV systems in the country to ensure the safe, reliable and ecient operation of the Electricity System. It includes governance; o-grid connections; grid connections; protection (to minimize damage to plant and consumer appliances); metering and information exchange requirements.

3.5 Standardisation of Solar PV System Components

ZERA is working with the Standards Association of Zimbabwe (SAZ) and other stakeholders to develop standards for solar PV system components such as batteries, panels, charge controllers, inverters, lighting kits and lanterns, system installation standards and for geysers. Once these standards are in place ZERA will enforce them through a Statutory Instrument on Solar PV regulations. In addition ZERA is set to fund the establishment of a dedicated solar PV equipment testing laboratory at the Standards Association of Zimbabwe to certify solar PV system components.

4. Zimbabwe Solar Energy Projects

The installation of solar photovoltaic systems in Zimbabwe dates back to the 1970s during which most of these systems were privately acquired. The Global Environmental Facility - United Nations Development Programme (GEF-UNDP) funded solar photovoltaic project that ran from 1993 to 1998, is the most popular project ever implemented by Zimbabwe. The thrust of the project was to address issues of the greenhouse gas emissions and global warming through the use of a sustainable model of solar electricity dissemination in rural areas.

The objectives of the project included the enhancement and upgrade of the country's solar manufacturing and delivery infrastructure through technical assistance, technician training, and provision of inputs to alleviate constrains on manufacturing. Over fifty (50) solar companies were active in the solar energy field by the time the project ended. The project delivered twelve thousand (12000) 45-watt equivalent solar home systems under subsidised conditions during the five year period.

The UNDP-GEF solar photovoltaic project is not the only solar project that the government embarked on. In 1992 – 1994, the German Technical Cooperation engaged into a USD 6.5 million solar water pumping project in Zimbabwe. The Japan International Cooperation Agency (JICA) also funded a solar project in 1997 – 1998. The JICA project focused on the installation of solar PV home systems. In addition, the Italian government at one time, ran a USD 92000 solar PV project that lit many rural schools. In 1999 the Chinese government also donated 110 solar PV systems and water pumps to rural communities.

The end of the GEF project saw the closure of solar companies that were active during the project due to lack of financial support. On the other hand, the credit mechanisms that were set to benefit the lower income groups in rural areas was also scraped. In addition, those that had benefited from the project faced problems relating to the maintenance of the installed solar home systems. Most people lost confidence in solar energy. In my opinion, the GEF project can be viewed as a failed Technology Transfer project. However, there were lessons learnt from this project that include the need to educate people to appreciate the advantages of solar energy and equip them with requisite skills for evaluation, design, installation and maintenance of solar systems.

Data gathered from the Department of Renewable Energy revealed that Practical Action – a development partner, installed a 100kW solar mini-grid in the Chipendeke in 2014 and Mashaba area of Gwanda in 2016. The solar mini-grid supplies electricity to a school, clinic, business centre and two irrigation schemes. Approximately 400 solar mini-grids were installed in rural schools and clinics by the Ministry of Energy and power Development each with a capacity of about 1kW in the last decade. 525 Solar PV mobile units have been distributed by the Ministry of Energy and Power Development to schools, police posts, national parks and health care centres in the remote areas of Zimbabwe. The Zimbabwe Power Company is working on three 100MW Solar Power Plants at Munyati, Gwanda and Insukamini. Currently, solar powered traffic lights and street lights are being installed by local municipalities.

Although an appreciable population has benefitted from the above-mentioned projects, the greater chunk of the population privately acquire solar systems. However, the challenges that people are facing are basically on the quality of the solar products that are imported into the country and the costs. The products that most people are getting are substandard and expensive.

5. Challenges

The current lack of a renewable energy policy, the fragmented nature of responsibilities for renewable energy development between government departments, the lack of a renewable energy feed-in tariff, and the general poor economic performance and lack of financial resources in the country, all contribute to the lack of development of sustainable energy sources. Below are some of the major challenges;

Cost

Solar energy technologies are still on the high side for the average Zimbabwean. Though efforts have been made by the government to reduce the cost on the end user by removing duty on solar systems components, the average cost is still high owing to the fact that Zimbabwe has no capacity to produce solar system components except solar batteries only. There is no capital to invest in large scale solar power generation. This is primarily because of low investor confidence on our Country. The Country recentlyadopted the use of bond notes due to lack of foreign currency. This affects importation of essential goods as well as pushing away investors. There is also a high political uncertainty attached to the country despite it being a safe destination for tourists.

Quality

At a lower level, there is a clear influx of unregistered and substandard solar components on the market. Most of the solar products that are imported by Zimbabweans are of low quality. There are high chances of failure of the solar systems in a short period of their use. In my own view the country is acting as a technology dump yard, receiving old technologies with poor performance.

Knowledge Barriers

The general population lacks knowledge of the design (evaluation of the size and system requirements), installation and maintenance of the solar systems. Lack of adequate knowledge may lead to improper usage and inability to maintain the solar system. This is the case with most Zimbabweans. The adoption of solar systems by people in rural communities is also affected by lack of knowledge hence the need for awareness campaigns. Installation of solar technologies is being done mostly by untrained and sometimes inexperienced people. Efforts are being made by the Government through ZERA to train as many technical people as possible. Technicians and installers are currently being trained countrywide to evaluate solar systems, install and repair them.

Regulatory Barriers

While Zimbabwe has set up a regulatory authority, it is not fully independent from state control, especially in determining tariffs. Proper pricing of power could help to ensure sustainable private sector participation.

6. Conclusions

As a Country, Zimbabwe needs to provide standards and strict regulatory measures to control the influx of substandard goods on the market. There is also need to provide incentives for investors in the form of repatriation of foreign funds, fiscal incentives to lower tariffs and capital costs for utility level projects, national project status among others. In addition, proper technology transfer frameworks would ensure an increase in the diffusion of solar technologies of better quality into the country.

REFERENCES

1. Bakhtyar, B., Zaharim, A., Asim, N., Sopian, K. and Lim, C.H., 2012, December. Renewable energy in Malaysia: Review on energy policies and economic growth. In *Proceedings of the 3rd International Conference on Development, Energy, Environment, Economics* (pp. 146-153).
2. Development of a National Integrated Energy Resource Plan for Zimbabwe – Terms of Reference, MOEPD, 2015, p. 1.
3. *Energy Efficiency and Renewable Energy Strategy Formulation for Zimbabwe*, by Francis Masawi, International Council of Swedish Industry, and Confederation of Zimbabwe Industries, Harare: 10 June 2013, p.13.
4. http://www.sunlightelectric.com/pvhistory.php
5. https://www.experience.com/alumnus/article?channel_id=energy_utilities and source_page=additional_articles and article_id=article_1130427780670
6. https://en.wikipedia.org/wiki/Timeline_of_solar_cells
7. Mulugetta Y, Nhete T, Jackson T. Photovoltaics in Zimbabwe: lessons from the GEF Solarproject.EnergyPolicy2000;28:1069–80. http://dx.doi.org/10.1016/S0301-4215(00)00093-8.

Chapter 19

Estimation of the Performance of a Solar Photovoltaic Power Plant Using Based Program

S. Das, T. Munetsiwa, V. Kundu, T. Kanyowa, E. Ndala, G. Kwinjo and G.V. Nyakujara

Renewable Energy Department,
Amity School of Applied Sciences (ASAS),
Amity University Haryana (AUH), Panchgaon,
Manesar-Gurugram – 122 413, Haryana, India
E-mail: sdas@ggn.amity.edu

The advent of solar technology has offered a clean, renewable and sustainable energy for the future. Enough data on solar radiation at any given location and at any given time of the year is critical for the design and installation of solar energy power plants. The need for integration of Solar PV power plants with the main distribution grid requires a more predictable and consistent power supply to meet the dynamic demand of domestic energy. The study presented in this paper highlights how the values of estimated solar radiation at any given time and location are imperative for determining the performance of photovoltaic solar power plant. It presents a software based tool that helps to quickly estimate the solar radiation for a given day throughout the year and determine the performance of any solar power plant based on the estimated data. This can be used by Design Engineers to study the feasibility of any solar power plant installation project. The estimation of solar radiation was done using the ASHRAE model and a software program platform (Visual Studio C#) was developed to model the overall performance of the power plant. The software aids in checking the feasibility of the installation of a solar photovoltaic power plant instantly, given the location and day of the year.

Keywords: *Solar radiation, ASHRAE Model, Photovoltaic power plant performance, Software.*

1. Introduction

Due to the environmental impacts caused by using non-renewable energy sources, there is a pressing need to speed-up the development of clean and renewable energy technologies. This has seen many governmental organisations enforcing renewable portfolio standards, requiring a certain percentage requirement of utility supplied power to come from renewable sources such as solar, wind, *etc.* Of all these renewable sources, solar energy is the most promising option due to its intrinsic benefits. The main attraction of the solar photovoltaic systems is that they produce electric power without harming the environment, by directly transforming the solar energy into electricity. The success of the solar photovoltaic system technology has resulted in concrete policy frameworks to promote and rapidly move towards the development of more solar power plant projects worldwide.

When considering to install a solar power plant at a location, it is of paramount importance to study the various factors that affect the performance of the system. This involves taking into consideration many factors such as the efficiencies of the PV panels that generate power, the regulators, the cabling and all the various components of the system. The weather conditions also influence the performance of the system which depends non-linearly on the solar insolation and temperature throughout the day and the year. Therefore a thorough analysis of these factors is imperative for the prediction of the performance and assessing the feasibility of a solar power plant project at a location. However, this can be a very tedious task to do manually.

The software presented in this paper provides a more convenient and systematic way to quickly predict the performance and feasibility of a proposed solar power plant project. It can be used to provide a comprehensive comparison between options of several solar photovoltaic systems that may differ with respect to design, technology or geographic location and finally validate models for system performance estimation during the design phase. It can also be used correctly manage power demand and energy planning at a specific location.

2. Modelling of Solar Radiation

performance analysis of any solar photovoltaic system involves evaluation of different instantaneous parameters that are recorded by the data acquisition system incorporated in the SPV system. The measurement of solar radiation over a period of time at a place is the best approach for estimating average radiation data for that place. In this study prediction of radiation; hourly global radiation and diffuse solar radiation falling on a horizontal surface was done using the ASHRAE model.

2.1.1 ASHRAE Model

The simplest approach to estimating solar radiation was developed by Souka and Safat [1] and later adopted or developed further by the American Society of Heating, Refrigeration and Air Conditioning (ASHRAE). The model gives equations based on an exponential decay model in which the beam radiation decreases with increase in the distance traversed through the atmosphere. [2]

The global radiation I_g reaching the horizontal surface on the earth is given by;

$$I_g = I_b + I_d \qquad (1)$$

where,

I_g: hourly global radiation

I_b: hourly beam radiation

I_d: hourly diffuse radiation

But $\mathbf{I_b = I_{bn} \cos \theta_z}$ (2)

I_{bn}: Beam radiation in the direction of the rays,

θ_z: angle of incidence on a horizontal surface *i.e.* the zenith angle. Therefore substituting

$$I_g = I_{bn} \cos \theta_z + I_d \qquad (3)$$

The ASHRAE model then postulates that;

$$I_{bn} = A \exp [-B/\cos \theta_z] \qquad (4)$$

$$I_d = CI_{bn} \qquad (5)$$

A, B and C are constants whose constants values have been determined on a north wise basis. These constants change during the year because of seasonal changes in the dust and water vapour contents as well as other aerosols in the atmosphere. The varying earth-sun distance also contributes to the changing of the constants values. The values of the A, B and C are given in the Table 19.1 and they are recommended for use. The values were initially suggested by Threlkeld and Jordan [3] and followed up for revision by Iqbal [4].

Table 19.1: Values of Constants A, B and C Used for Predicting Hourly Solar Radiation on Clear Days

	A (W/m^2)	**B**	**C**
January	1202	0.141	0.103
February	1187	0.142	0.104
March	1164	0.149	0.109
April	1130	0.164	0.120
May	1106	0.177	0.130
June	1092	0.185	0.137
July	1093	0.186	0.138
August	1107	0.182	0.134
September	1136	0.165	0.121
October	1136	0.152	0.111
November	1190	0.144	0.106
December	1204	0.141	0.103

The ASHRAE model is a simplified model which takes an integrated view of the attenuation process rather than considering the attenuation caused by individual constituents.

2.1.2 Performance Estimation of a Solar Photovoltaic Power plant

The amount of energy or power output by a solar PV system or power plant is largely dependent on the following factors;

- ☆ Size or capacity of the power plant
- ☆ Amount of solar irradiation received or sun peak hours
- ☆ Performance degradation over life cycle
- ☆ Total efficiency of considering all the losses

To calculate the energy yield that a power plant can produce we can express it mathematically in the form;

Energy yield = Peak sun hours Array rated power
Total De-rating/Loss Factor [6]

It should be noted that the Peak sun hours is the solar energy available at a given location expressed as KW-h/m^2/day. The array rated power is simply the capacity of the power plant and the de-rating factor is nothing but the efficiency after incorporating all losses

2.2 Components of a Solar Power Plant

A typical Solar PV power plant, is a power system designed to supply usable solar power by means of photovoltaics. It consists of an arrangement of several components, including solar panels to absorb and convert sunlight into electricity, a solar inverter to change the electric current from DC to AC, as well as mounting, cabling and other electrical accessories to set up a working system [4]. It may also use a solar tracking system to improve the system's overall performance and include an integrated battery solution, as prices for storage devices are expected to decline. Strictly speaking, the array of solar panels, the visible part of the PV system is the most important part of the system, it can produce electricity (DC) without the use of auxiliary hardware components, though it is of primary importance to protect the system (by making use of fuses and proper cables) and to convert the Direct current to Alternating current so as to suite the electrical devices available commercially in the market.

2.3 Losses Associated with a PV Power Plant

The losses in a PV are caused by several factors and they reduce the overall performance and efficiency of a Solar Power Plant.

2.3.1 Losses Associated with a PV Module

The incident solar radiation on the PV module is not entirely converted to useful energy. Some of it is lost. The first energy loss occurs in the Solar Cell and this can be due to the fundamental reason (limited by material property) or it could be due

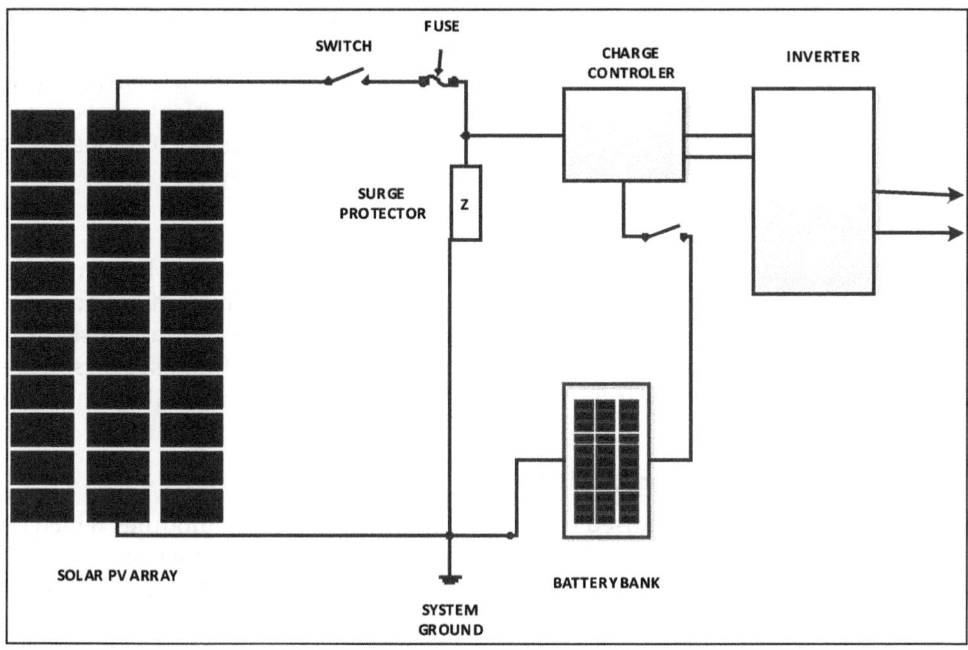

Figure 19.1: The Major Components of a Solar Photovoltaic Power Plant.

to the technological reason (limited by processing capabilities). Photon energy loss can occur due to the following factors:

- ☆ Transmission losses
- ☆ Thermalization losses
- ☆ Fill factor losses
- ☆ Optical Losses [5]

2.3.2 Losses Due to Electrical Conductors

These are ohmic losses and they are dependent on the sizes of the wire conductors used in the system. These losses can minimized by optimizing cable sizing and cable routing so that the losses are less than 1.5 per cent. [6]

2.3.3 Losses Due to Temperature

This aspect is more important in countries with relatively higher temperature such as India where the temperatures can go as high as 50 degrees Celsius. The power output of a typical module reduces by about 0.42 per cent to 0.48 per cent for every degrees Celsius rise in cell temperature. For example for a cell temperature of 70 degrees Celsius, a 230Wp module may only produce about 185W. [7]

2.3.4 DC to AC Conversion Losses (Inverter Efficiency)

These losses are due to the presence of a transformer and the associated magnetic and copper losses, inverter self-consumption, and losses in the power electronics. The conversion efficiency is defined as the ratio of the fundamental component of the AC power output from the inverter, divided by the DC power input. The conversion efficiency is not constant, but depends on the DC power input, the operating voltage, and the weather conditions including ambient temperature and irradiance. The variance in irradiance during a day causes fluctuations in the power output and Maximum Power Point (MPP) of a PV array. [8]

2.3.5 Soiling

Soiling occurs as a result of dust and dirt accumulation on Solar panels. In some cases, the material may be washed off the panel surface by rainfall; however dirt like bird droppings may stay even after heavy rains. The dirt causes shading of the cells and reduces the available power from a module. In most cases, these losses account for 1 per cent loss; however the power can be restored if the modules are cleaned. [9]

2.3.6 Losses Due to Array Mismatch

Mismatch losses are caused by the interconnection of solar modules which do not have identical properties or under different conditions in series and parallel. Mismatch losses causes the output of the entire PV array to be determined by the solar module with the lowest output. The array is formed by connecting the modules in a series parallel configuration to match the input DC voltage of the inverter. Generally, modules are specified with a tolerance of +/-2 per cent to +/-5 per cent (or 0 per cent to +5 per cent) [10]

3. Results and Discussion

Figure 19.2 shows the software algorithm for the program developed. It commences with the input data such as latitude of the place, date of the year, rated power, losses of the power plant and plant capacity. Arithmetical operations are then performed based on the input data and given formulas mentioned in the above equations. The displayed output is the global radiation, beam radiation and diffuse component of radiation which further aids in giving the estimated output of the power plant.

Figure 19.3 shows the interface of the software or rather the home page when you log into the system. Figure 19.4 shows the performance tab with the display inputs but without entered inputs. Figure 19.5 then depicts the performance tab after inputs have been entered and an analysis graph of the global radiation is generated. Therefore depending on the location, day of the time and expected losses and capacity of the power plant, the estimated output can be generated at any given day and location.

4. Conclusions

This application is quite useful especially during the pre-installation period where quick decisions need to be made based on numerical estimates of data at the possible least amount of cost. This tool for estimating the Solar PV power plant performance is beneficial in reducing the time to assess project feasibility and it

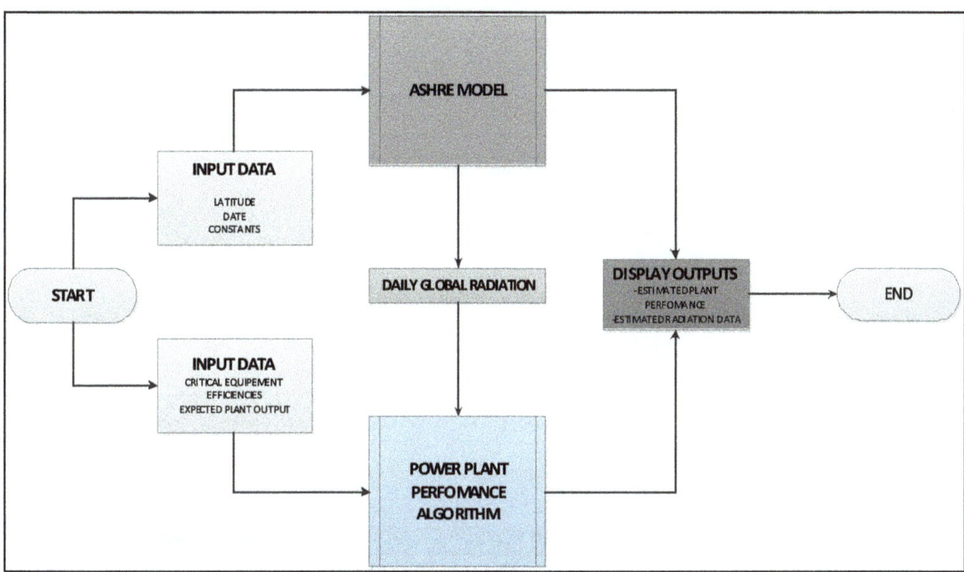

Figure 19.2: Software Algorithm.

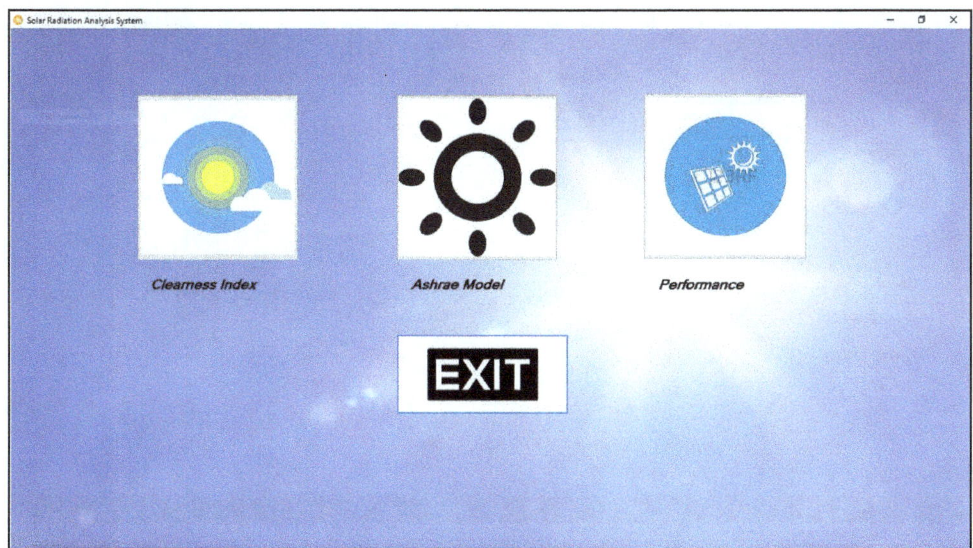

Figure 19.3: Interface of the Software, Home Page.

simplifies the design process of a PV plant. It helps to evaluate whether the output of a Solar PV plant will meet the required load so that proper decisions can be made as to whether a backup and/or auxiliary power plant is needed. Design Engineers can use this tool to quickly compare the performance of a Solar Photovoltaic plant at several locations with different conditions.

Performance Estimation
SPV Performance Estimation Software
Latitude ϕ: 25.1525
Date 26
Rated power kW;
Plant Losses %
Module Loss %: 0
Cables Loss %: 0
Invertor Loss %: 0
Charge Controller %: 0
Plant Losses %
Battery Loss %: 0
Accessories Loss %: 0
Year of Operation: 0
Peak Sun Hours
Derating factor %:
Expected Output (kWh/day):
Hourly Global Radiation
GlobalRad
Calculate
Exit

Figure 19.4: Performance Tab without Inputs.

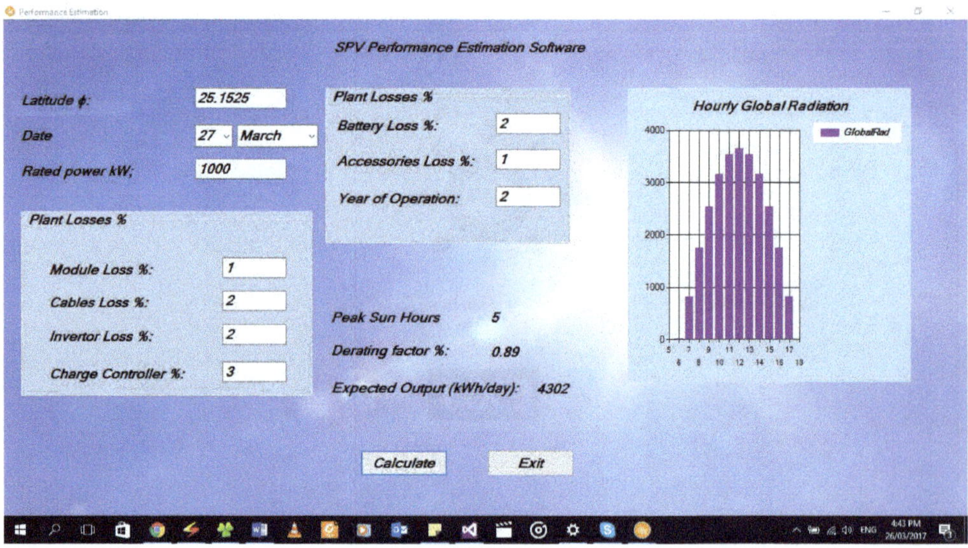

Figure 19.5: Performance Tab with Inputs and Analysis Graph.

REFERENCES

1. Souka A.F., Safwat H.H., Determindation of the optimum orientations for the double exposure flat-plate collector and its reflections, Solar Energy vol.10, pp 170-174. 1966.

2. ASHRAE 1972. Handbook of Fundamentals, American Society of Heating, Refrigeration and Air-conditioning Engineers, pp 385-443
3. Threkeld, J.L and Jordan, R.C. 1985. Direct solar radiation available on clear days. ASHARAE Transactions, 64: 45.
4. Pearce, Joshua (2002). "Photovoltaics – A Path to Sustainable Futures". Futures. **34** (7): 663–674
5. Green, M.A. Solar Cells: Operating Principles, Technology and System Applications, Prentice Hall Inc. Englewood Cliffs, N.J, USA 1982
6. http://www.re-solve.in/perspectives-and-insights/solar-pv-plant-performance-capacity-utilisation-factorcuf- per cent 20/(accessed 8/9/2013).
7. http://www.schatzlab.org/docs/PVArrayMismatchLosses.pdf/(accessed 4/9/2014).
8. http://techno.su.lt/~bielskis/straipsniai per cent 20ir per cent 20knygos/Theoretical per cent 20assessment per cent 20of per cent 20the per cent 20maximum per cent 20power per cent 20point per cent 20tracking.pdf (accessed 5/7/2014).
9. http://www.civicsolar.com/resource/effect-array-tilt-angle-energy-output (accessed 8/8/2013).
10. http://files.sma.de/dl/7680/Perfratio-UEN100810.pdf/(accessed 2/8/2013

Chapter 20

The Possibility of Sustainable Renewable Energy for Nigeria as the Way Forward

M.M. Gaji[1] and A.B. Mohammed[2]*

[1]Energy Commission of Nigeria, Abuja, Nigeria
[2]Nigerian Electricity Regulatory Commission, Abuja, Nigeria
**E-mail: engrmmgaji@yahoo.co.uk*

This country report analyzes various works reviewing the renewable energy potential of Nigeria. It raises the possibility of having Nigeria's electricity grid powered by 100 per cent renewable energy at the earliest possible date. At present, only 10 per cent of rural households and 30 per cent of the country's total population have access to electricity (the third largest country without access to electricity according to the IEA). Nigeria has one of the lowest net electricity generation per capita rates in the world: 15,000MW peak load demand to 6,050MW available capacity, resulting in an unreliable power supply and a correspondingly heavy dependence on fossil fuels in industries and residential areas.

The majority of electricity generation in Nigeria came from fossil fuels (79 per cent) at the moment, with about two-thirds of thermal power derived from natural gas and the rest from oil, resulting in CO2 emission growth. Nigeria currently ranks 46th in the world for CO2 emissions released, with over 73.69 metric tons released in 2011 (World Bank). With global climate change prevailing, shifting to renewables (which would reduce the CO2 emissions) would be the best rational option to salvage our degrading environment whilst providing a quality power supply to the tens of millions of people in rural areas currently without access to electricity. Harnessing energy from wind and the sun would further help reduce poverty and improve standard of living.

Technical research conducted recently concludes that a 100 per cent stable power supply from renewable energy is possible in Nigeria. Nigeria is positioned perfectly at the moment for investment in renewables, especially as it continues to fight for a stable power supply with several power plants currently under construction and the privatization of the power sector from November, 2013. The feasibility and technicalities are elaborated across the chapters in this paper.

1. Overview of Nigeria

Nigeria lies in the western region of Africa and is boarded by The Gulf of Guinea to the South, Niger to the North, Cameroon to the East and Benin Republic to the West. Nigeria's land area is 923,768 km^2. Nigeria is the most populous country in Africa with about 170 million people. Due to Nigeria's prime location along the equator, the climatic condition throughout the year is generally favorable, but varies.

Nigeria is also endowed with huge deposits of oil, natural gas and other natural resources. It is the largest oil producer in Africa as well as has the largest gas reserve (Ajayi, O.O, 2013). Petroleum has been the mainstay of Nigeria's economy since its discovery (Akinbami, J.F.K., 2001).

Figure 20.1: Map of Nigeria Showing Boundaries. Google 2013 Map Data.

It plays a significant role in the nation's international diplomacy and it serves as a tradable commodity for earning the national income, which is used to support government development programmes. It also serves as an input into the production of goods and services in the nation's industry, transport, agriculture, health and education sectors, as well as an instrument for politics, security and diplomacy.

The energy sub-sector, especially petroleum, continues to maintain its prominence as the single most important source of government revenue and foreign exchange earner. Petroleum contributed an average 25.24 per cent to the GDP between 2002 and 2006. However, despite the fortunes of the oil sector, other sectors of the economy are declining. For example, consumption of electricity actually declined by 13.4 per cent between 2002 and 2006 even though the overall or total electricity consumption showed a marginal increase of 1.8 per cent from 5.63GWh in 2002 to 7.47GWh in 2006. Only about 40 per cent of households in

Nigeria are connected to the national grid. There is high-energy loss due to the physical deterioration of the transmission and distribution facilities, an inadequate metering system and an increase in the incidence of power theft through illegal connections. Other problems of the power sector include manpower constraints and inadequate support facilities, the high cost of electricity production, inadequate basic industries to service the power sector, poor billing systems, poor settlements of bills by consumers and low available capacity, about 40 per cent out of the installed capacity of about 6,000MW. Inadequate funding prevented targeted growth in the sector. Production activities in the solid minerals sub-sector were generally on decline. The situation in the rural areas of the country is that most end users depend on fuel wood. The contribution of energy to GDP is expected to be higher when we take into account renewable energy utilization, which constitutes about 90 per cent of the energy used by the rural population (NPC, 1997).

Despite the large abundance of fossil fuel deposits in Nigeria, it also has immense potential in the renewable energy field; especially in hydropower generation because of its prime location with access to 840 km coastline in the South and two great rivers entering from the Northeast and Northwest. The northern part of Nigeria is also very close to the Sahara, hence enough sunlight and medium winds for substantial power generation.

1.2 Current Energy Consumption

Energy Information Administration, EIA estimates that in 2011, Nigeria's primary energy consumption was about 4.3 Quadrillion Btu (111,000 kilotons of oil equivalent) (Asiegbu, A.D and Iwuoha, 2007). Of this, traditional biomass and waste accounted for 83 per cent of total energy consumption. This high per cent represents the use of biomass to meet off-the-grid cooking heating and cooking needs, mainly in rural areas. Nigeria has vast natural gas, coal and renewable energy resources that could be used for domestic electricity generation, yet lacks policies to harness resources and develop new (and improve current) electricity infrastructure.

1.3 Energy Situation in Nigeria

The country is located between longitude 8 E and latitude 10 N, and has two major seasons, wet and dry. The seasonality makes water availability at the different hydropower stations variable, leading to an intermittent supply at times of low water levels. Also, the thermal power stations have been bedeviled by lack of adequate supplies of natural gas from the various Niger Delta gas wells, thereby making continuous energy production from these installations difficult (ECN and UNDP, 2005). This has left Nigerians at the mercy of private, alternative power generation through the use of diesel and petrol generators (Esan A.A., 2003).

Nigeria's electricity sector is relatively small. Brazil and Pakistan, two countries with similar population sizes, generate 24 times and 5 times more power than Nigeria, respectively (Fadare, D. A., 2008). The latest EIA estimates show that Nigeria's net generation was 18.8 billion kilowatt-hours (KWh) in 2009 (Hoogwijk, *et al.*, 2003). The graph below clearly illustrates the abysmal population to power ratio in Nigeria (5GW to 170 million people) (Iwayemi, A., 2008).

Figure 20.2: Nigeria's Power Grid (*Source*: Google).

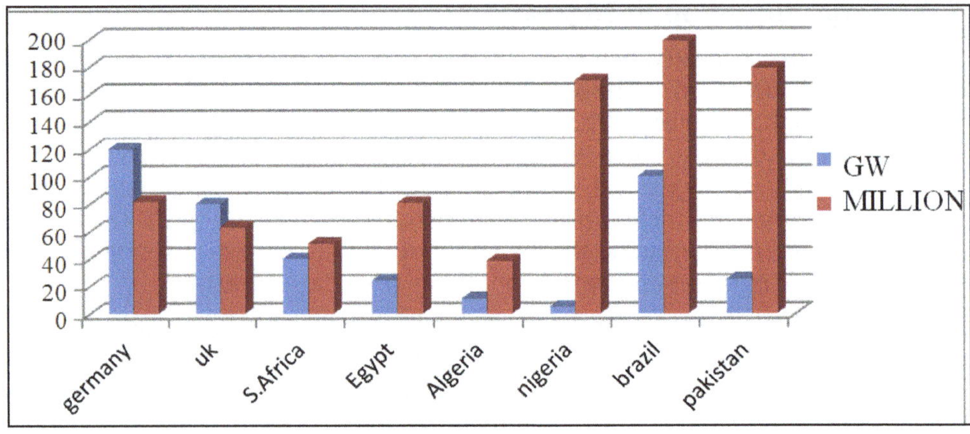

Figure 20.3: Countries Population to Generation Capacity Ratios.

The installed power capacity has remained relatively flat over the last decade 10,396.0 MW, although net energy generation has slightly decreased from its peak of 23 billion KWh in 2004, mainly due to a decline in hydroelectric power (IEA, 2013). The majority of electricity generation in Nigeria comes from thermal power plants (79 per cent), with about two-thirds of thermal power being derived from natural gas and the rest from oil. Hydroelectricity (21 per cent), the only other source of power generation, has decreased from its peak of 8.2 billion kWh in 2009, slightly less than generation, and exported most of the remainder to Niger through an agreement under the West African Power Pool (Ismaila Haliru Zarma, 2006). IEA data for 2009 indicates that the electrification rate for Nigeria was 50 per cent for the country as a whole - leaving approximately 80 million people without access to electricity in Nigeria (KPMG Report, 2013). Other estimates place the countrywide electrification rate as low as 45 per cent.

Fuel wood is used by over 60 per cent of Nigerians living in the rural areas. Nigeria consumes over 50 million metric tonnes of fuel wood annually, a rate, which exceeds the replenishment rate through various afforestation programmes. Sourcing fuel wood for domestic and commercial uses is a major cause of desertification in the arid-zone states and erosion in the southern part of the country. The rate of deforestation is about 350,000 hectares per year, which is equivalent to 3.6 per cent of the present area of forests and woodlands, whereas reforestation is only at about 10 per cent of the deforestation rate (Inter-Ministerial Committee report, 2000). The rural areas, which are generally inaccessible due to absence of good road networks, have little access to conventional energy such as electricity and petroleum products. Petroleum products such as kerosene and gasoline are purchased in the rural areas at prices 150 per cent in excess of their official pump prices. The daily needs of the rural populace for heat energy are, therefore, met almost entirely from fuel wood.

Rural electrification is given high priority in government's efforts to increase the standard of living in rural areas, reduce rural-urban migration trends, and realize other development objectives. However, the three key challenges for rural electrification are:

1. How to provide sustainable energy (electricity) services to the poorest of the poor, who have no purchasing power to pay for the services?
2. How to offer the most cost-effective, clean and reliable electricity to those who are currently spending a significant share of their income on energy?
3. How to set up the commercial infrastructure to provide these services?

Solar power generation which seemed neglected is now receiving huge attention, forming the center of many power generation-focused conversations in Nigeria. Power problems in the country mirror that of many energy markets in Africa. Nigeria's power ministry estimates that about 40,000 MW of electric power is required to satisfy the country's industrial demand but Nigeria only has about 12,000 MW of installed generation capacity, of which around 3,500 MW are supplied by gas-fired plants and hydro power systems. The majority of the gas-fired power plants are now burdened by problematic gas supplies, arising from pipeline vandalism by militant groups operating in the gas producing region of the Niger Delta.

The Renewable energy program of the Nigerian government says "about 600,000 MW of electricity can be generated from just one per cent of Nigeria's land mass." This implies that Nigeria can procure 100 per cent of its power from a renewable source such as solar. The government is now poised and willing to take advantage of its solar resource endowment by supporting investments in utility-scale solar power projects. This ambition helped spur the recently signed Power Purchase Agreement (PPA) with 14 utility-sized solar power developers, which will add about 1,200 MW of solar capacity to the grid.

Solar power investors are also optimistic about the Nigerian opportunity. Justin Woodward is the Chief Development Officer and Co-Founder of Canadian JCM Capital, which is one of the 14 solar power developers now operating in Nigeria. Mr. Woodward believes "Nigeria has the potential to lead in utility-scale solar power in Africa, given its well-structured regulatory frameworks, the standard of which is mostly absent in many other African countries. From the good structure of the bulk trading company to the transmission company and the independent Distribution Companies (Discos), all of which combine to provide a level of operational clarity those investors required." JCM Capital has recently commenced the development of an 80 MW solar plant in the Nigerian northern state of Katsina.

This renewed interest in solar power does not interest only utility-scale projects but also mini-grid solar power plants. The Nigerian government plans to generate about 2000 MW of electricity from renewable sources by 2020, according to the power ministry. In addition, government is exploring a competitive tendering process that will attract more solar power investors.

Despite the appetite for solar power investment being still high, recent macroeconomic crisis in Nigeria could become a hurdle that may limit the realization of the country's potential. Mr. Woodward believes that Nigeria's economic recession could slow the pace of further solar investments and with respect to the proposed tendering process he says that "unless the economy improves in the medium term, investors may not be willing to negotiate lower than the present tariff in a bidding exercise."

For Nigeria's renewable mini-grids systems generating less than 30 MW, there will be a dedicated new regulation, which came into place in 2015, known as the Renewable Energy Feed-in-Tariff (REFIT). The benefits for projects that function within the REFIT framework includes: market stability through guaranteed long term power pricing schemes, quick access to the grid, streamlined licensing procedures and government's obligation to purchase power generated by the renewable plants.

The Ministry of power says REFIT is designed to "reduce the transaction costs associated with negotiating and signing a PPA for a small renewable generator (and) the feed-in-tariff is accompanied by standardized PPAs for projects of up to 30 MW". Improving the regulations that govern solar power projects in Nigeria has further enhanced that country's attractiveness to investors and the achievements to be recorded in terms of ongoing and planned projects is expected to have more knock-on effects, since Nigeria's solar irradiation is already enormous and the required policies are increasingly being put in place.

However, challenging areas for solar power developers still exist in Nigeria. Some developers and observers have mentioned the poor state of transmission and distribution infrastructures as a major barrier to development and a source of significant risks to investors. Mr. Woodward says "while accelerated investment and regulations can increase solar power generation, the poor wheeling capacity of the transmission infrastructure in the country is still a major constraint." In actuality, Nigeria's transmission infrastructure can only handle 6000 MW in its current state.

The successful completion and operation of these ongoing projects is expected to motivate some established utilities in Nigeria (who at the moment generate power with little or no renewable resources) to consider the incremental inclusion of technologies such as solar power into their generation mix. This will help fast track Nigeria's journey to the status of a nation that predominantly uses clean energy and a leader in solar power development within the Sub-Saharan Africa region.

2. Solar Energy Potential

The sun is the most readily and widely available renewable energy source capable of meeting the energy needs of whole world (J. O. Oji, *et al.*, 2012). One of the greatest assets that Nigeria has that can facilitate solar energy generation in Nigeria is her geographical location, that is, the equatorial region which is full of large quantity of solar radiation. Solar radiation is fairly well distributed in Nigeria with an average solar radiation of about 19.8 $MJ/m^2/day$ and average sunshine hours of 6 hours a day; ranging between about 3.5 hrs at the coastal areas and 9.0 hrs at the far northern boundary (D. Abdulsalam, *et al.*, 2012). Nigeria lies within a high sunshine belt and thus has enormous solar energy potentials. If solar collectors or modules were used to cover 1 per cent of Nigeria's land area, it would be possible to generate 1850 103 GWh of solar electricity per year; this is over one hundred times the current grid electricity consumption level in the country (Titilayo A. Kuku). Generally, the solar capacity for Nigeria ranges between 3.5kW/m/day – 7.0kW/m/day and average sunshine daily of 4 – 7 hours (Ikechukwu Ikeagwuani, et al).

In Nigeria, more than 70 per cent of the population is rural dwellers. Since the energy production level of any community dictates her pace of development, it is possible to alleviate poverty of the large community of Nigerians by providing alternative renewable energy (solar) for them. Several pilot projects, surveys and studies have been undertaken by the Sokoto Energy Research Center (SERC) and the National Center for Energy Research and Development (NCERD) under the supervision of the Energy Commission of Nigeria (ECN). Several PV-water pumping, electrification, and solar-thermal installations have been put in place. Such solar thermal applications include solar cooking, solar crop drying, solar incubators and solar chick brooding. Other areas of application of solar electricity include low and medium power application such as: water pumping, village electrification, rural clinic and schools power supply, vaccine refrigeration, traffic lighting and lighting of road signs.

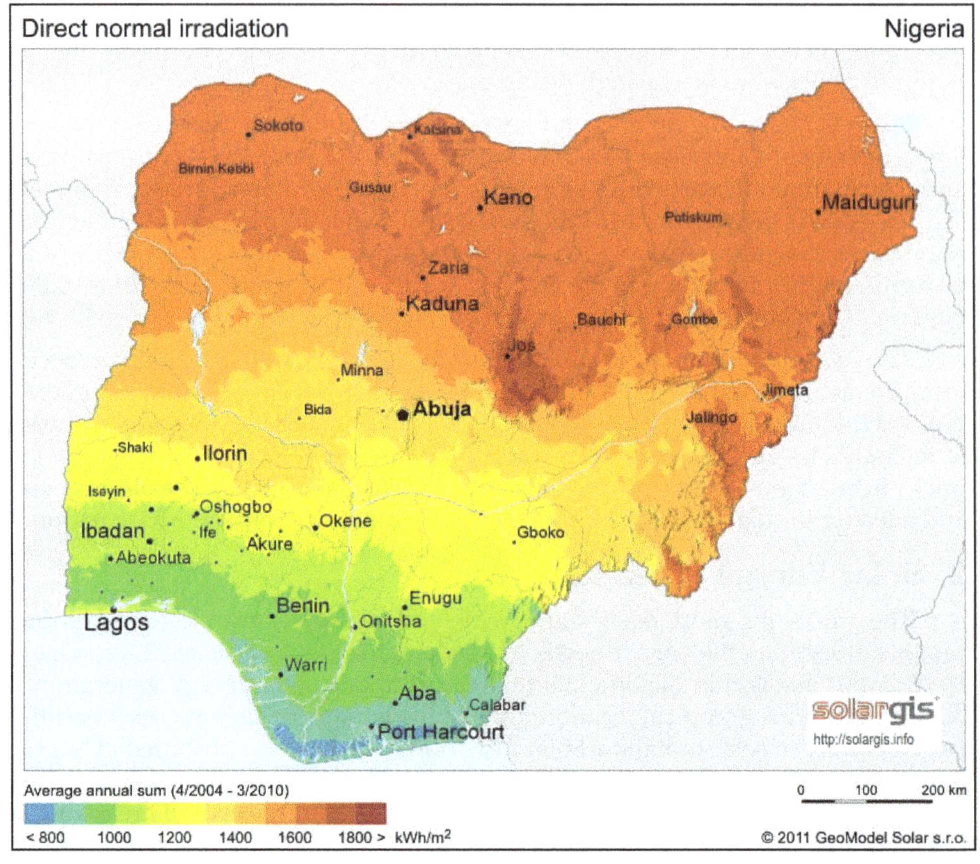

Figure 20.4: Nigeria's Average Solar Radiation Map.

2.1 Solar Power Technologies

Solar energy can be used to generate power in two ways; solar-thermal conversion and solar electric (photovoltaic) conversion.

2.1.1 Solar-thermal Conversion

Solar-thermal is the heating of fluids to produce steam to drive turbines for large-scale centralized generation. Like solar cells, solar thermal systems, also called concentrated solar power (CSP), use solar energy to produce electricity, but in a different way. Most solar thermal systems use a solar collector with a mirrored surface to focus sunlight onto a receiver that heats a liquid. The super-heated liquid is used to make steam to produce electricity in the same way that coal plants do.

2.1.2

The Renewable Electricity Action Program (REAP) of the Federal Ministry of Power and Steel (2006) published by the International Centre for Energy,

Environment and Development in Nigeria did not cover this aspect of power generation.

2.1.3 Solar Electric (Photovoltaic) Conversion

Solar-electric (photovoltaic) conversion is the direct conversion of sunlight into electricity through a photocell. This could be in a centralized or decentralized fashion. Solar-electric (Photovoltaic) technologies convert sunlight directly into electrical power. Photovoltaic system is made up of a balance of system (BOS), which consists of mounting structures for modules, power conditioning equipment, tracking structures, concentrator systems and storage devices. Photovoltaic conversion could be small scale for stand-alone systems or large scale connected to the national grid.

Solar cells are also referred to as photovoltaic (PV) cells, which as the name implies (photo meaning "light" and voltaic meaning "electricity"), convert sunlight directly into electricity. Panel stands for a group of modules are connected mechanically and electrically. A module is a group of cells connected electrically and packaged into a frame (more commonly known as a solar panel), which can then be grouped into larger solar arrays. Photovoltaic cells are made of special materials called semiconductors, such as silicon which is most commonly used. When light strikes the cell, a certain portion of it is absorbed within the semiconductor material and the energy of the absorbed light is transferred to the semiconductor. The energy knocks electrons loose, allowing them to flow freely (SciTech, 2014).

Nigeria lies within a high sunshine belt and thus has enormous solar energy potentials. The mean annual average of total solar radiation varies from about 3.5 kWhm^2/day in the coastal latitudes to about 7kWhm^2/day along the semi arid areas in the far North. On the average, the country receives solar radiation at the level of about 19.8 MJ/m^2/day. Average sunshine hours are estimated at 6hrs per day. Solar radiation is fairly well distributed. The minimum average is about 3.55 kWhm^2/day in Katsina in January and 3.4 kWhm^2/day for Calabar in August and the maximum average is 8.0 kWhm^2/day for Nguru in May.

Given an average solar radiation level of about 5.5 kWhm2/day, and the prevailing efficiencies of commercial solar-electric generators, then if solar collectors or modules were used to cover 1 per cent of Nigeria's land area of 923,773km^2, it is possible to generate 1850x103 GWh of solar electricity per year. This is over one hundred times the current grid electricity consumption level in the country.

Solar thermal applications, for which technologies are already developed in Nigeria, include: solar cooking, solar water heating for industries, hospitals and households, solar evaporative cooling, solar crop drying, solar incubators and solar chick brooding.

Solar electricity may be used for power supply to remote villages and locations not connected to the national grid. It may also be used to generate power for feeding into the national grid. Other areas of application of solar electricity include low and medium power application such as: water pumping, village electrification, rural clinic and schools power supply, vaccine refrigeration, traffic lighting and lighting of road signs, *etc.* Several pilot projects, surveys and studies have been undertaken by the Sokoto Energy Research Center (SERC) and the National Center for Energy Research and Development (NCERD) under the supervision of the Energy Commission of Nigeria. Several PV-water pumping, electrification, and solar-thermal installations have been put in place.

2.3 Estimated Resource Base

The National Energy Policy Document states that "Nigeria lies within a high sunshine belt and, within the country; solar radiation is fairly well distributed (A.A. Adeyanju, 2011). The annual average of total solar radiation is varies from about 12.6 $MJ/m^2/day$ (3.5 $kWh/m^2/day$) in the coastal latitudes to about 25.2$MJ/m^2/day$ (7.0 $kWh/m^2/day$) in the far north."(J. Radiol Prot, 2009).

Thus, over a whole year, an average of 6,372,613 PJ/year (dikka1,770 thousand TWh/year) of solar energy falls on the entire land area of Nigeria. This is about 120 thousand times the total electrical energy generated by the National Electric Power Authority for the whole country for the year 2002 (P.A. Nwofe, 2014). This then is an estimated potential solar thermal energy resource base.

Part of this resource falls on agricultural and forest lands and of course is useful for photosynthesis processes; others fall on developed areas, which could be harnessed for power generation through roof- and building-integrated designs. There could be immediate and widespread deployment of solar energy which could easily cover a large area of Nigeria, especially rural and riverine areas. For rural areas there are many advantages of solar energy over the present energy supply sources, especially a decentralized application. The decentralization of solar energy installation means individual acquisition, utilization and application of the system. No high or low tension transformer would be required, high or low tension wiring, equipmentand logistic would be involved in the distribution of the energy, which means the solar PV panels could easily be carried, deployed and installed on individual establishment and premises in any part of the country at low cost within a very short period of time.

3. Problems Confronting Solar Installation in Nigeria

3.1. Affordability

Nigeria is still classified as underdeveloped country with high percentage of her population living under poverty level. This makes the ability to acquire solar energy devices, which are still considered expensive, difficult for both individuals and groups.

3.2. Government Ignorance

The people in government who make policy are very much unaware of the capacity of solar energy; many people assume that solar energy can power only small bulbs or at most television set. The print media also has not produced enough publicity on the subject matter. It is evident that the problems of solar energy in Nigeria, though enormous, but can be addressed within short period if government were to give proper attention to research, development, commercialization and installation of solar equipment through good policy evolution

3.3. Component Failure

Since the process of solar energy is very new in this part of the world, users get turned off if it does perform up to the years of guarantee which the equipment is rated. Equipment and component failure occurs mostly with components that do not carry a manufacturer's address or guarantee.

3.4. Cost of Generation

Comparing equipment and installation cost of solar energy with other energy supply sources, solar energy is more expensive in the short run. However, it is cheaper in the long run. The results show that the PV source is more expensive up to 4 years after installation. This is because solar energy components are very expensive and they are mostly imported except the cables and few accessories. Beyond 5 years, PV power becomes more attractive because of low running cost. A higher percentage of the population in Nigerian is low income earners and cannot afford or acquire solar power. However, higher income earners have access to other energy sources like petrol/diesel generators apart from the grid.

3.5. Political Problems

The politics behind acquisition and installation of solar energy at both the governmental and technical level in Nigeria are not encouraging. On the government's part, there is no clear cut legislation backing the utilization of renewable energy. The government has not at any time embarked on giant step by installing or acquiring large solar power plants.

3.6. Identification of Good Geographical Location

The problem of geography can be solved by integrating the most relevant and important aspects of solar energy installation and generation into the curriculum of the professional training schools, such as the civil and surveying, who are the first contact point for proponents and users of solar energy.

Communiqué reached in the 2007 National Energy Forum (NASEF), among others, identified additional challenges to solar energy development in Nigeria as: cultural restriction on land use, lack of appropriate institutional framework, low level of technical expertise, vandalization and theft of system components and lack of local manufacturing of system component-PV.

3.7. Lack of Focus on the Renewable Energy Master Plan

Nigeria's Renewable Energy Master Plan of 2005 (ECN-UNDP) says that the country should endeavor to increase the energy generation capacity from 5000 MW to 16000 MW by 2015 through the exploration of renewable energy resources. At present, there has not been a single grid generation of electricity from renewable energy sources (besides the traditional hydropower generation prequel to the REMP) probably as a result of government's lack of focus and commitment to the plan. The government at all levels needs to be committed to a plan they have initiated and agreed to if there must be meaningful development. The Renewable Energy Master Plan will be a vital resource if there can be serious devotion to the suggestions contained therein. Part of this suggestion includes the suspension of the Renewable Energy (RE) import duties, integration of RE into non-energy sector policies, establishment of national RE development agency, standardization of RE products and establishment of RE fund to provide incentives, micro-credits schemes, training and also funding R&D (A. Charles, 2014). Moreover, there may be a need for the master plan to be broken down into renewable source components, with each addressing expected contributions from particular type of renewable resources.

3.8. Lack of Adequate Funding

Lack of adequate funding has been a major setback in the growth of WET and other renewable energy technology in Nigeria. Annually, the percentage of the federal budget to education and science and technology ministries has not been encouraging. With the meager sum made available to these ministries, much productive research and development may not be started or supported. Corporate bodies also need to be encouraged to collaborate with research institutions to fund research aimed at national development, some of which include wind-for- power projects (both small and medium scale turbines), nationwide wind energy resource assessment, development of adequate and explanatory national wind atlas/map that would provide information on quantity, distribution, quality and utilization possibilities to determine the commercial feasibility of wind energy generation and decision making on investment and development of national wind turbine tests and certification (E. Hackett, 2010).

3.9. Other Challenges

Apart from all the earlier mentioned challenges, which if overcome will move the nation forward in utilizing wind for power generation, there are other challenges which include lack of awareness and technical ineptitude. The level of awareness on the viability of wind power is very low in the country. The majority of schools' curriculum lack adequate information on wind and other renewable resources. Mass media too has not helped in any way, hardly any information regarding wind energy utilization or technology may be seen on the pages of newspaper or heard discussed on television or radio. This lack of awareness has also led to high level technical ineptitude, thereby making adoption of wind as veritable source of power generation a difficulty (O. O. Ajayi, 2010).

4. Conclusions

Renewable energy is considered a viable solution to the energy challenges of Nigeria especially in the rural areas of the country and to the restrictions posed by the rising cost of conventional or traditional energy. In this article, the role of renewable energy technologies in meeting the energy challenges is discussed. Also consideration has been given to the factors affecting developments in the renewable energy sector, and efforts made to ensure capacity building for renewable energy, stimulation of the private sector, developing the markets for renewable energy, obtaining the necessary finance for renewable energy projects and the assistance of multilateral institutions in advancing renewable energy technologies in the country.

REFERENCES

1. Ajayi, O.O, "Modelingthewind energypotential ofNigeria", Covenant University, Ota, Ogun State, Nigeria, 2007, Retrieved July2013.
2. Akinbami, J.F.K. (2001),"RenewableEnergyResourcesand Technologiesin Nigeria: Present Situation, Future Prospects and PolicyFramework. Mitigation andAdaptation StrategiesforGlobal Change", 6: 155-188, Netherlands, Retrieved August,2013.
3. National Planning Commission (1997), "National Rolling Plan (1997 – 1999).
4. Report of the Inter-Ministerial Committee on Combating Deforestation and Desertification,August 2000.
5. Asiegbu,A.D andIwuoha, "G.S. Studies of wind resources in Umudike, South East Nigeria– Anassessmentof economicviability," Journal of Engineering and Applied sciences 2(10), 2007, 1539–1541, Retrieved July2013.
6. ECN and UNDP, Nov. 2005. "NigeriaRenewableEnergyMaster Plan(REMP)"; Retrieved October, 2013.
7. www.areanet.org/fileadmin/user_uploada/AREA/AREA_downloads/AREA_Conferene Presentations/Nigeria_Renewable_Energy_Masterplan.pdf>
8. Esan A.A., "Preparedness on Development GreenE nergy for Rura lIncome Generation- Nigeria's Country Paper UNIDO, INSHP/IC SHP," Hangzhou, China June 19-23, 2003, Retrieved August, 2013.
9. Fadare, D.A. "A Statistical analysisof wind energypotential inIbadan, Nigeriabased on Weibull distribution function",Thepacific journalof science and technology,9(1), 2008, 110–119, Retrieved July2013.
10. Hoogwijk, M.,FaaijA.,van denBroek R.,Berndes G., Gieten, D., TurkenbergW. (2003), "Exploration ofthe ranges of theglobal potential ofbiomass for energy, Biomass and Energy", Vol. 25 No. 2; 119– 133, RetrievedAugust, 2013.
11. Iwayemi, A., Nigeria's dual energyproblems: Policyissuesand challenges, International Association for energyeconomics, Fourth Quarter, 2008, available online www.iaee.org/en/publications/newsletterdl.aspx?id=53, Accessed 21 December2009, 17–21 Association for energyeconomics,FourthQuarter, 2008
12. International EnergyAgency IEA, "Nigeria", Accessed August 2013, www.iea.org/countries/non-membercountries/nigeria
13. IsmailaHaliruZarma, "Hydro Power Resourcesin Nigeria, ECN, 2006,"http://www.unido.org/fileadmin/import/52413_Mr._Ismaila_Haliru_Zarma.pdf, Retrieved August, 2013.KPMG 2013 Report, "Oil and Gas in Africa"; Retrieved November, 2013.
14. www.kpmg.com/africa/en/issuesandinsights/articles-publications/pages/oil-gas-in- africa.aspx.

15. J. O. Oji1, N. Idusuyi, T. O. Aliu, M. O. Petinrin, O. A. Odejobi, A. R. Adetunji, 2012, "Utilization of Solar Energy for Power Generation in Nigeria" International Journal of Energy Engineering 2012, 2(2): 54-59. DOI: 10.5923/j.ijee.20120202.07
16. D. Abdulsalam, I. Mbamali, M. Mamman, Y. M. Saleh, 2012, "An Assessment of Solar Radiation Patterns for Sustainable Implementation of Solar Home Systems in Nigeria" American International Journal of Contemporary Research, Vol. 2 No. 6; June 2012
17. Titilayo A. Kuku, 2008, "Improved Availability of Clean Non Grid Energy in Nigeria Through the Use of Renewable Energy Sources and Energy Efficiency Principles", Department of Electronic and Electrical Engineering,Obafemi Awolowo University,Ile-Ife, Nigeria
18. [1]Ikechukwu Ikeagwuani, [2]Olusola Bamisile, [3]Serkan Abbasoglu, [4]Arua Julius, "Performance Comparison of PV and Wind Farm in Four Different Regions of Nigeria". [1,2,3]Department of Energy Systems Engineering, Cyprus International University, Haspolat-Lefkosa, Via Mersin 10, Turkey. [4]Department of Mechatronics Engineering, Akanu Ibiam Federal Polytechnic, Unwana.
19. SciTech, 2014, "Sunlight knocks electrons loose from solar cell material" The tartan.
20. A.A. Adeyanju, 2011, "Solar Thermal Energy Technologies in Nigeria"Research Journal of Applied Sciences, 2011 | Volume: 6 | Issue: 7 | Page No.: 451-456, DOI:10.3923/rjasci.2011.451.456.
21. J. Radiol Prot, 2009, "Human exposure to high natural background radiation: what can it teach us about radiation risks?" Published online 2009 May 19. DOI: 10.1088/0952-4746/29/2A/S03.
22. P.A. Nwofe, 2014, "Utilization of Solar and Biomass Energy- A Panacea to Energy Sustainability in a Developing Economy", International Journal of Energy and Environmental Research Vol.2, No.3,PP.10-19, September 2014. Published by European Centre for Research Training and Development UK (www.eajournals.org).
23. A. Charles, 2014, "Global Energy Network Institute (GENI)". www.geni.org.
24. O. O. Ajayi, 2010, "The Potential for Wind Energy in Nigeria". Wind Engineering Volume 34, No. 3, 2010, Multi-Science Publishing Company.
25. Hackett, E. (2010), "Peer Review and Social Science Research Funding" in World Social Science Report 2010, Paris, France: UNESCO.

Dubai Resolution on

Trends in Solar Power Generation and Energy Harvesting in Developing Countries

WHILE EXPRESSING GRATITUDE to the National Science Centre, Ministry of Science, WHILE EXPRESSING GRATITUDE to the Centre for Science and Technology of the Non-Aligned and Other Developing Countries (NAM S&T Centre) for organising the on 'Science Centres in Promoting A Knowledge and Innovative Society for SustaInternational Workshop on **'Trends in Solar Power Generation and Energy Harvesting'** at Dubai during 27–29 March 2017.

EXPRESSING APPRECIATION to the Amity University, UP, India – Dubai Campus for co-organising and hosting the International Workshop.

RECOGNISING that solar power, especially as it reaches more competitive levels with other renewable energy sources, may serve to sustain the lives and families of millions of underprivileged peoples in developing countries.

HAVING CONSIDERED that renewable energy resources can improve quality of life by promoting sustainable development and systems such as solar power are practical, reliable, cost-effective, and healthier for people and the environment.

HAVING DELIBERATED on how solar power is the best choice for sustainability and renewable energy in developing countries, and how completed projects and on-going work in remote locations may create confidence to take up more ambitious projects.

TAKING INTO ACCOUNT that presently there is an urgent need for efficient harnessing of various forms of solar energy *i.e.* Solar Thermal and Photovoltaic in GW capacities with efficient technologies demonstrated for lightning needs, potable water generation and distribution, mid-day meal preparation and distribution in schools, vaccine preservation, health care, information distribution and comfort needs of aging population in NAM and other developing countries.

WE, THE PARTICIPANTS OF THE WORKSHOP, comprising Scientists, Researchers, Engineers and Policy Makers representing the governments, institutions and agencies from **Afghanistan, Cambodia, Cuba, Egypt, The Gambia, India, Indonesia, Iran, Iraq, Malaysia, Mauritius, Morocco, Nepal, Nigeria, Palestine, South Africa, Sri Lanka, Tanzania, Togo, Turkey, Zambia and Zimbabwe.**

UNANIMOUSLY RESOLVE AND RECOMMEND THAT:

- ☆ The governments in the developing countries should undertake appropriate policy measures to promote renewable energy development including solar PV and Solar thermal technologies and applications, appropriate for local conditions of economic development, social development and resources availability.
- ☆ Strong enabling policy framework should be evolved to promote an environment where renewable energy technologies do not face unfair disadvantages compared to conventional energy technologies to nurture the local markets.
- ☆ Adequate support mechanisms should be provided by the Governments to promote Solar Energy in order to fulfil its potential and contribute to meeting the energy needs in the developing countries.
- ☆ Governments in the developing countries should steer resources mobilised for large-scale investments into new production sectors and new technologies. The policies should base on active industrial policies, combining large scale investments and active policy interventions.
- ☆ Governmental incentives should be provided to strengthen the solar PV market that may positively affect installations for developing nations. More funding will make solar energy economical in on-grid markets, which will then lower the prices because of the high volume of manufacturing.
- ☆ New and innovative implementation models to finance and operate solar energy technologies are needed, to overcome the barrier of high capital costs and encourage the widespread use of renewable energy in rural and remote communities.
- ☆ Programmes on awareness creation and capacity building for planners and other relevant agencies should be initiated.
- ☆ The skills needed to deliver the Solar Energy programme in its entirety, including the long term support required for appropriate training for the service delivery chain need to be analysed thoroughly.
- ☆ The informal, traditional, micro, small and medium scale manufacturing sectors for solar energy devices should be transformed to become environmentally and financially sustainable by fostering R&D, capacity building and skill upgradation so that they are globally competitive.
- ☆ The trends in Solar Power Generation and Energy Harvesting are presently crystalline c-Si based. In this area manufacturing in developing countries is currently insignificant as compared to the global output for proliferation of solar power generation through PV route. Therefore, they will have to

solely depend on import of PV panels. As such supporting R&D in this area in developing countries, with sub critical funding, without producing poly Silicon material cheaply and in quantities, may not necessarily translate into lower cost of solar cell manufacturing. It is therefore suggested that a task force of NAM and other developing countries may be formed to look into this problem and suggest remedial action.

- ☆ Solar thermal power generation appears to be attractive in the above context in these countries. However its proliferation will need massive investment to compete with solar PV based power generation. Techno-economics of this initiative, therefore, needs to be closely worked out.
- ☆ As cooling is the most pressing demand to maintain productivity in most of the NAM countries, efficient ways of solar cooling systems of various capacities need to be consciously developed and deployed. This requires to be studied in all its details and necessary blueprint for R&D leading to manufacturing of such systems be prepared.
- ☆ Availability of potable drinking water purified through RO, desalination and other emerging technologies utilising solar energy in schools and other remote communities, should be implemented.
- ☆ Scheffler type dishes and other related technologies for cooking mid-day meals for school children using solar thermal energy may be encouraged.
- ☆ Standardization and certification mechanisms for solar equipment should be strengthened.

Considering the high quality research infrastructure and expertise available in the Amity University and other institutions, in the area of Solar Energy, it was suggested that the scientists and researchers from the developing countries may be given opportunities for short term affiliation in such institutions for their capacity building in this field. The detailed terms and conditions for such arrangements may be settled after mutual consultation.

To facilitate close interaction and information dissemination among scientists of various countries, a website has been created by Amity University, which can lead to many collaborative programmes and also help in developing training programmes. The website has been launched as: www.amitynamstworkshop.com.

THUS RESOLVED AND ADOPTED AT DUBAI, UAE THIS DAY, THE 29TH OF MARCH, TWO THOUSAND SEVENTEEN.

Fig. 1.1: Most Solar Panels in the Kingdom are Part of Solar Home Systems or SHS (*Courtsey*: SEAC). P (4)

Fig. 1.2: The Solar Farm of the New Coca-Cola Plant in the Phnom Penh Special Economic Zone Supplies Roughly of Third of its Electrical Consumption (*Courtesy*: Coca-Cola)/ P (5)

Fig. 1.4: The PV Panels on the Rooftop of Silvertown Metropolitan will Provide Approximately 12 per cent of the Building's Energy Needs. P (9)

Fig. 2.1: PV Module. P (16)

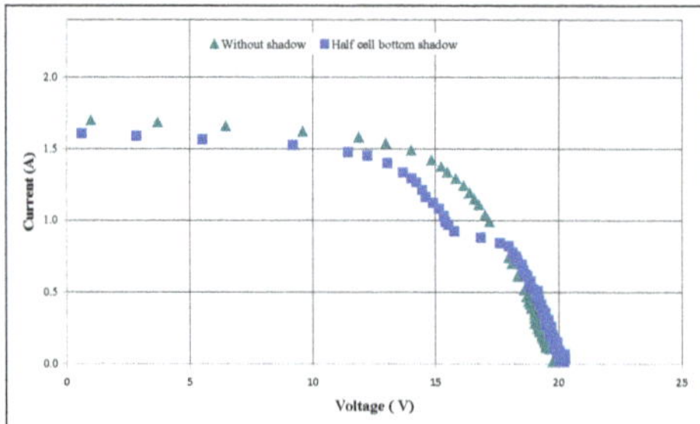

Fig. 2.3: I-V Characteristics of the PV Module for a Bottom Half Cell Shaded. P. (19)

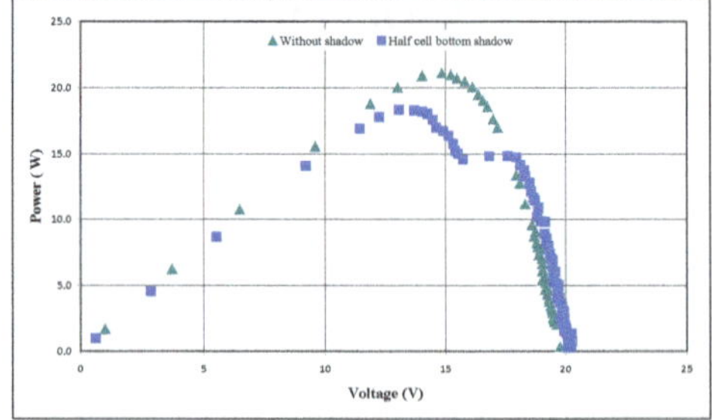

Figuer 2.4: P-V Characteristics of the PV Module for a Bottom Half Cell Shaded. P (19)

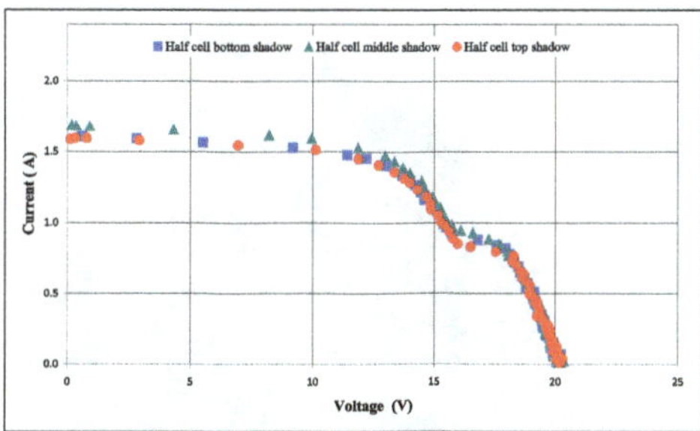

Fig. 2.5: I-V Characteristics of the PV Module for a Bottom, Middleand Top Half Cell Shaded. P (20)

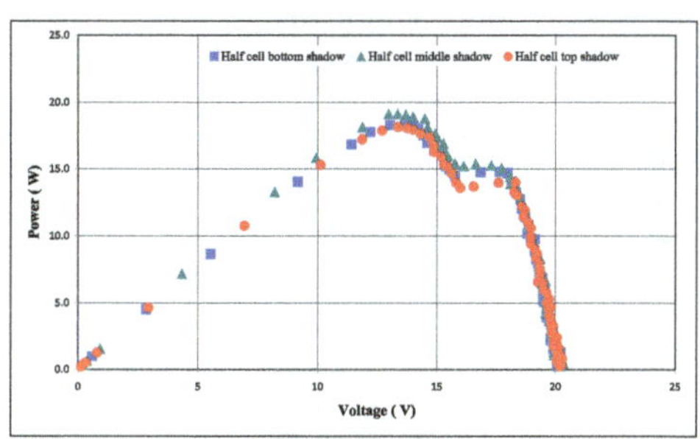

Fig. 2.6. P-V Characteristics of the PV Module for a Bottom, Middleand Top Half Cell Shaded. P (20)

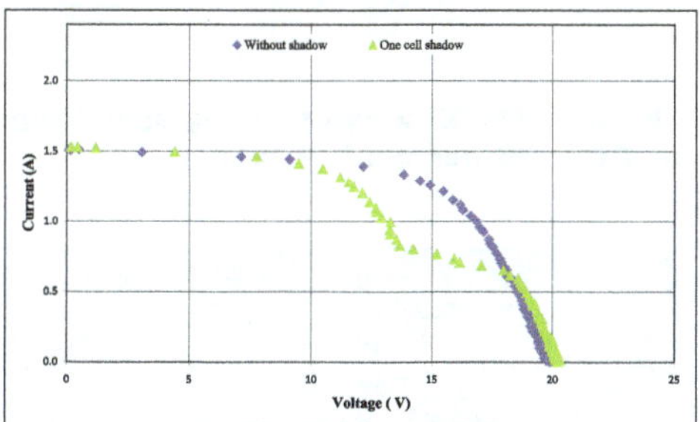

Fig. 2.7: I-V Characteristics of the PV Module for One Cell Shaded. P (21)

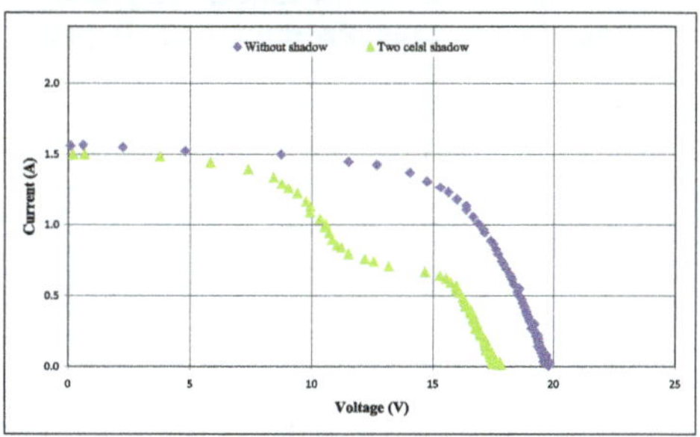

Fig. 2.8: I-V Characteristics of the PV Module for Two Cell Shaded. P (21)

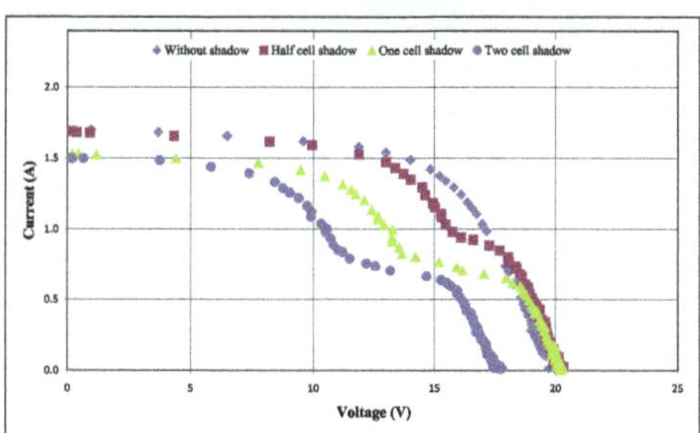

Fig. 2.9: I-V Characteristics of the PV Module for Half Cell, One Cell and Two Cells Shaded. P (22)

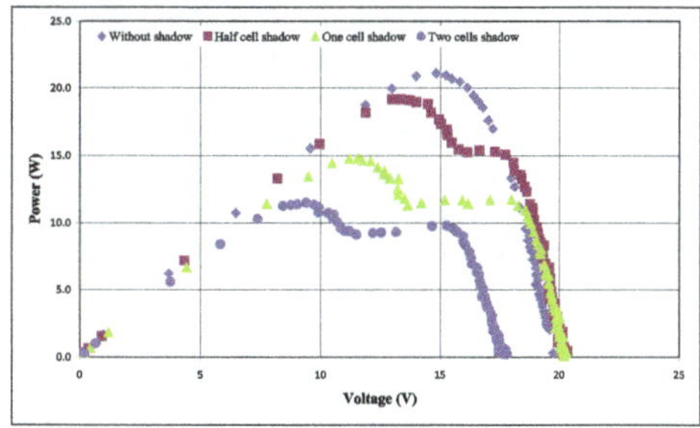

Fig. 2.10: P-V Characteristics of the PV Module for Half Cell, One Cell and Two Cells Shaded. P (23)

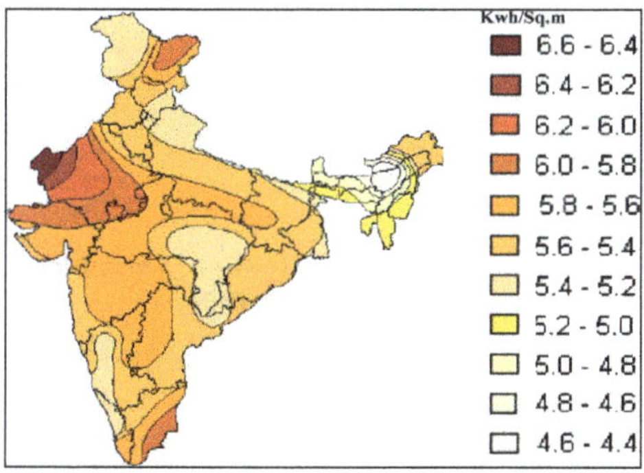

Fig. 3.1: Solar Radiation on India. P (30)

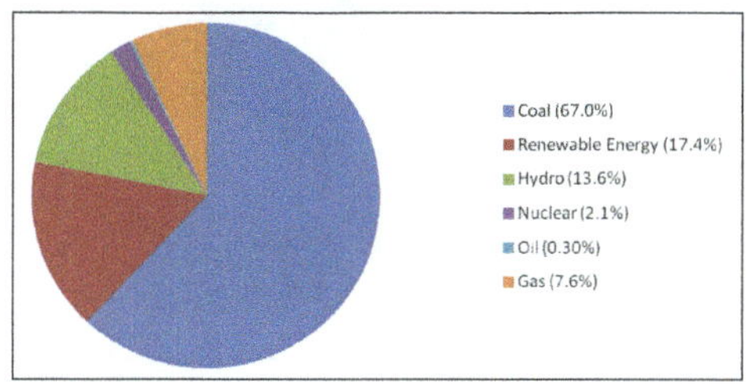

Fig. 3.2: Total Installed Power Generation Capacity of India (As on 30 June 2017). P (30)

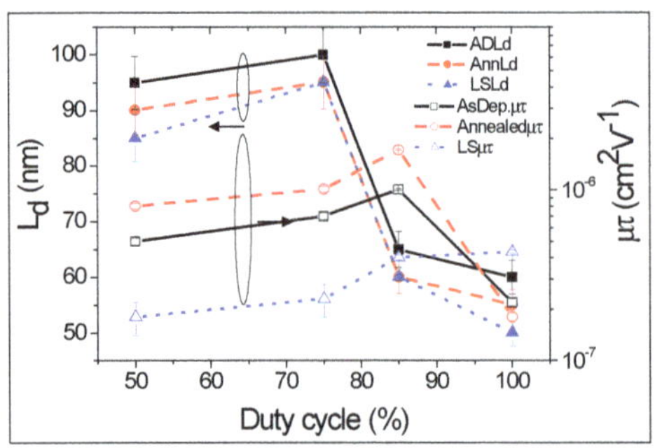

Fig. 4.1: Light Induced Degradation of Ambipolar Diffusion length (L_d) and Mobility Lifetime Product (μt). P (41)

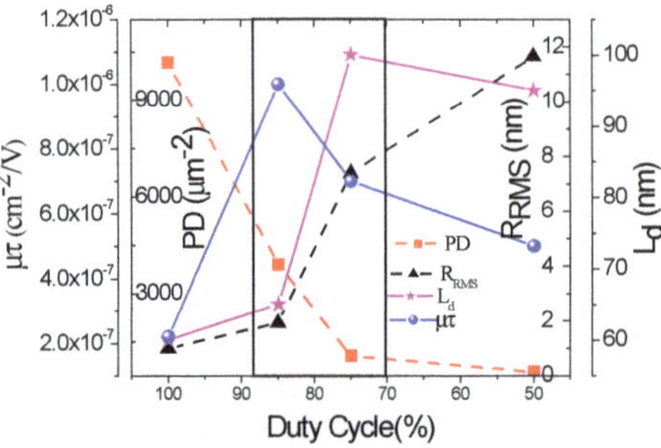

Fig. 4.5: Optimization of the Film Quality in the Samples by Control of Particle Size and of the Ion Bombardment by the Variation in the Duty Cycle. The shaded region in between the two vertical dotted lines is the region for optimum deposition conditions. P (44)

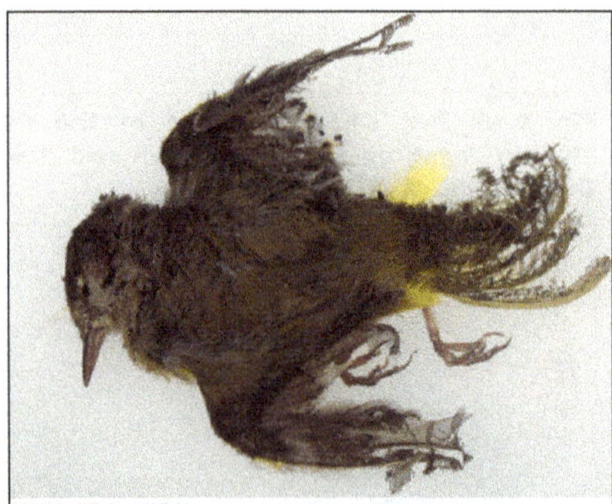

Fig. 5.2: A burned MacGillivray's Warbler that was Found at the Ivanpah Solar Plant during a Visit by U.S. Fish and Wildlife Service in October 2013. P (49)

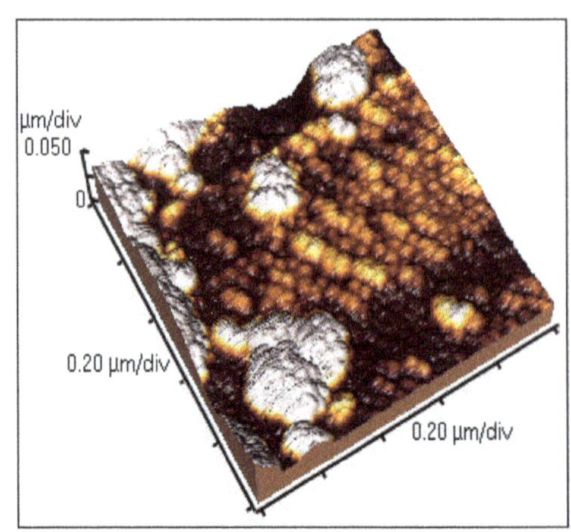

Fig. 4.3: AFM 3D Surface Topography Image of DC = 75 per cent. P (43)

Fig. 5.1: Ivanpah Solar Thermal Power Plant. P (48)

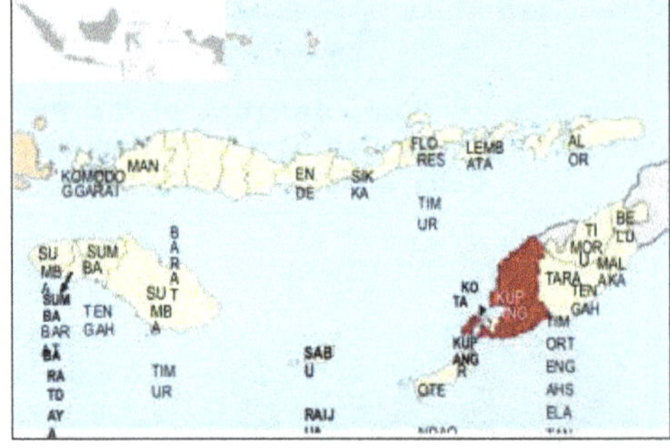

Fig. 6.1: Map of East Nusa Tenggara Province. P (59)

Fig. 6.2: Area of Kupang Regency. P (60)

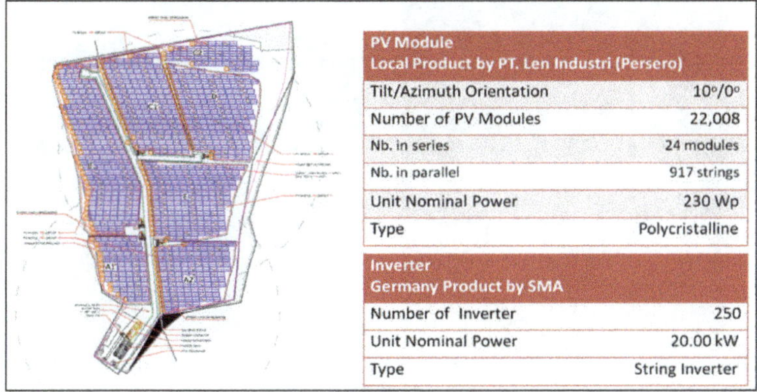

PV Module Local Product by PT. Len Industri (Persero)	
Tilt/Azimuth Orientation	10°/0°
Number of PV Modules	22,008
Nb. in series	24 modules
Nb. in parallel	917 strings
Unit Nominal Power	230 Wp
Type	Polycristalline

Inverter Germany Product by SMA	
Number of Inverter	250
Unit Nominal Power	20.00 kW
Type	String Inverter

Fig. 6.4: Site Layout of Photovoltaic Generator. P (62)

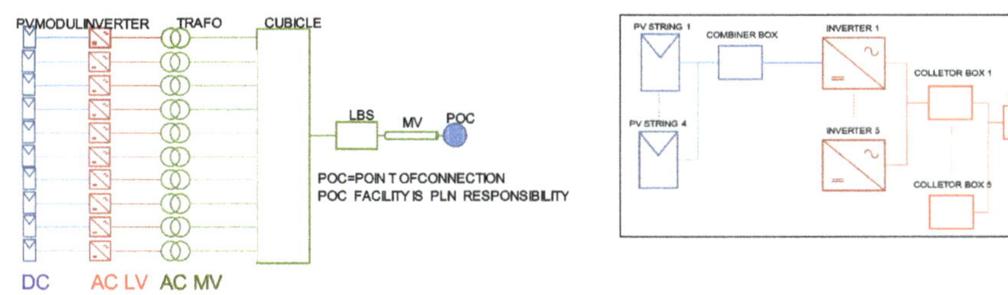

Fig. 6.5: Block Diagram of Photovoltaic Generator. P (63)

Fig. 6.6: Detail Block Diagram of DC and Low Voltage AC Sides. P (64)

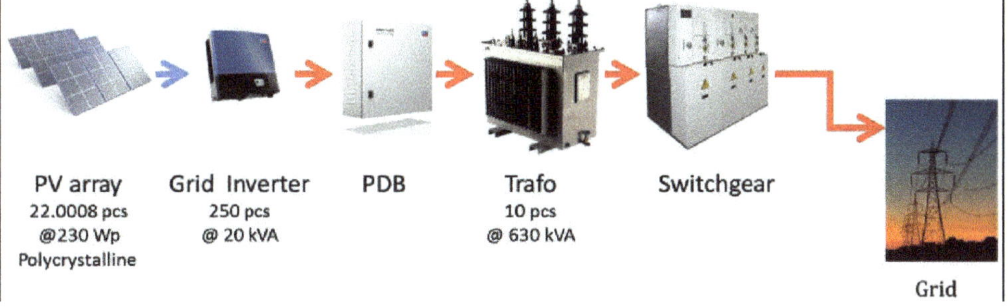

Fig. 6.7: Power Plant Configuration. P (64)

Fig. 6.8: Eelectricity Grid from the Power Plant to the Nearest Medium Voltage grid of PLN. P (64)

Fig. 6.9: Aerial Picture of Photovoltaicarray in Kupang. P (65)

Fig. 6.10: Inauguration of the Power Plant by the President of the Republic Indonesia. P (66)

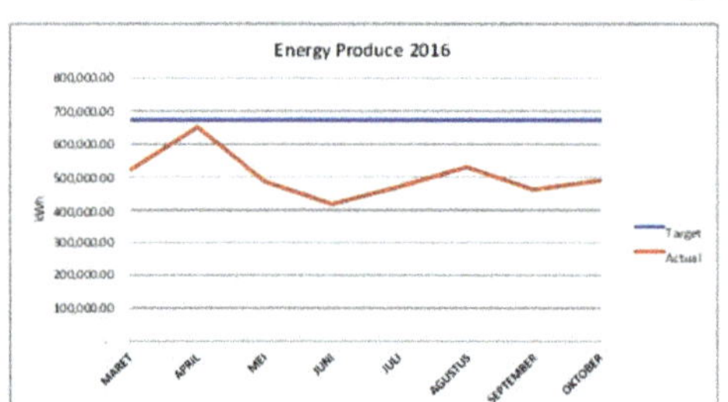

Fig. 6.13: Energy Produce Always Below the Energy Targeted. P. (67)

Power (kW) dan Frequency (Hz)

Power

Frequency

Fig. 6.14: Full System Operation on November 20th 2016. P (68)

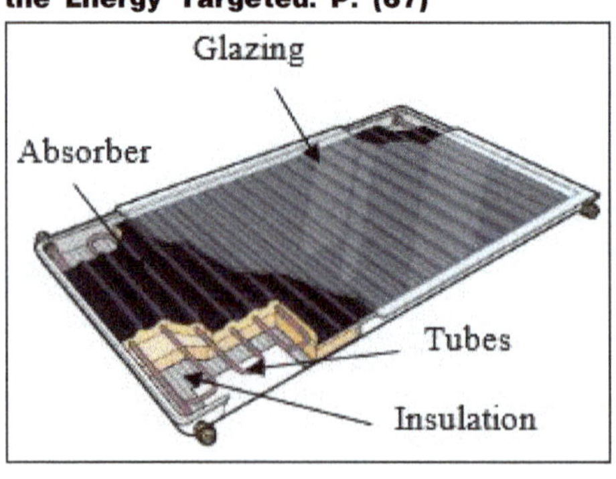

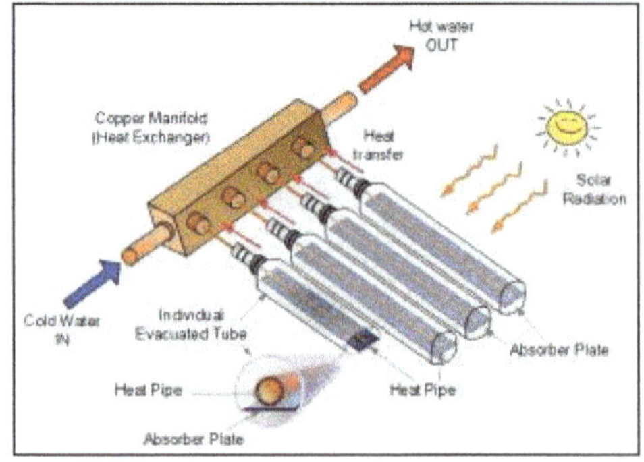

Fig. 7.1: Construction of Flat Plate Collector (Left) (Trust, 2015) and Evacuated Tube Collector (Right) (Tutorials, 2015). P (74)

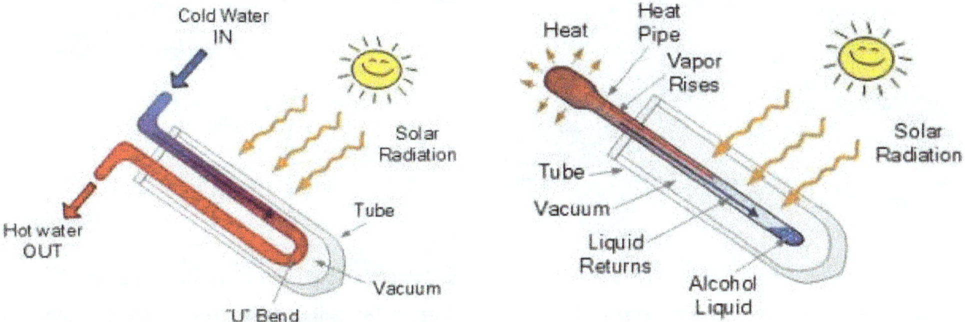

Fig. 7.2: Direct Flow (Left) and Heat Pipe Evacuated Tube Collector (Right) (Tutorials, 2015). P (75)

Fig. 7.8: 746.9 kW_{th} Solar Thermal Plant for Meat Industry in Austria (INTEC, 2015). P (81)

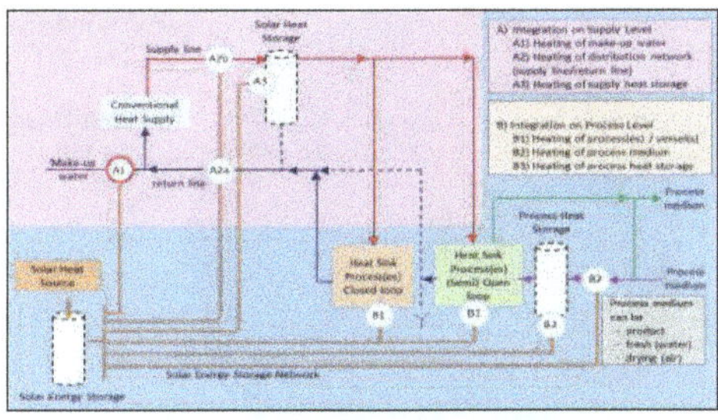

Fig. 7.9: Solar Heat Integration Point for Fleischwaren Berger Plant (INTEC, 2015). P (82)

Fig. 7.10: 1,064 kW_{th} Solar Thermal Plant for Beverage Industry in Austria (INTEC, 2015). P (82)

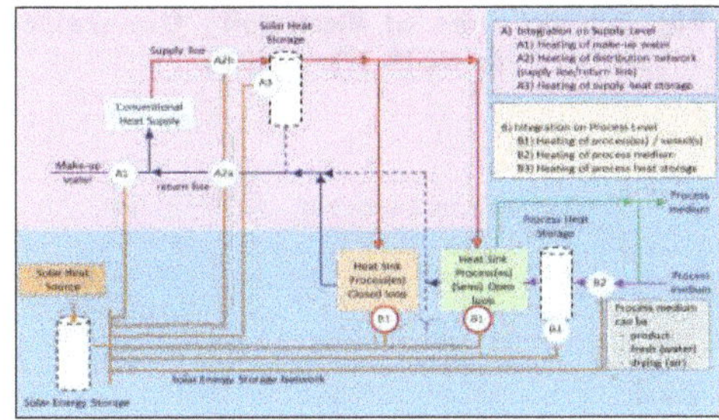

Fig. 7.11: Solar Heat Integration Point for Goess Brewery Plant (INTEC, 2015). P (83)

Fig. 7.12: 83 kW_{th} Solar Thermal Plant for PPNJ Poultry and Meat industry in Malaysia. P (83)

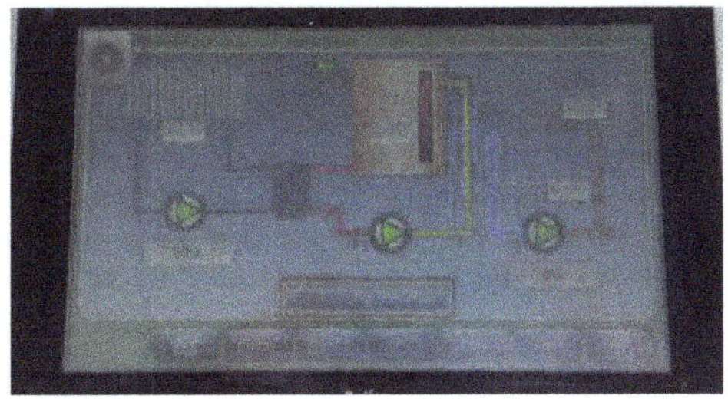

Fig. 7.13: Solar Heat Integration Point for PPNJ Poultry and Meat, Malaysia. P (84)

Fig. 7.14: 31.7 kW_{th} Solar Thermal Plant for De Baron Resort Hotel, Langkawi, Malaysia. P (85)

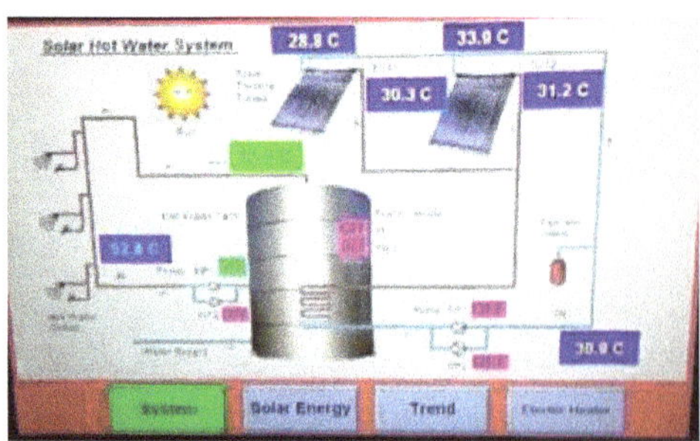

Fig. 7.15: Solar Heat Integration Point for De Baron Resort Hotel, Malaysia. P (85)

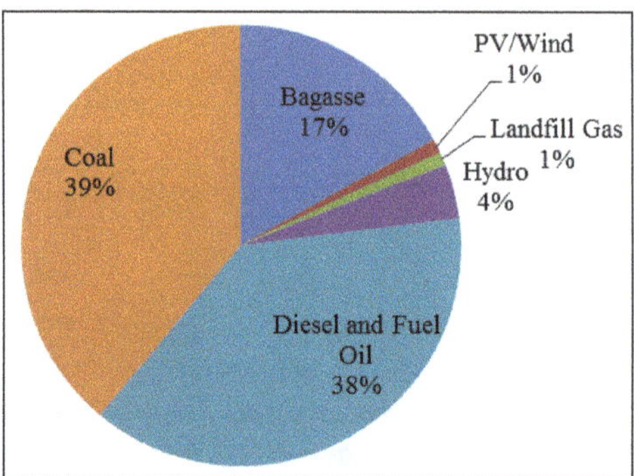

Fig. 8.1: Sources of Electricity Generation in 2015 [1]. P (90)

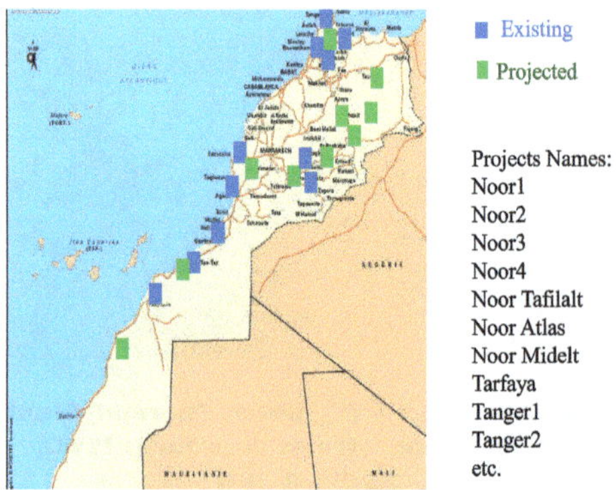

Fig. 9.2: Solar and Wind Large Energy Projects. Existing (in blue) and projected (in green). P (106)

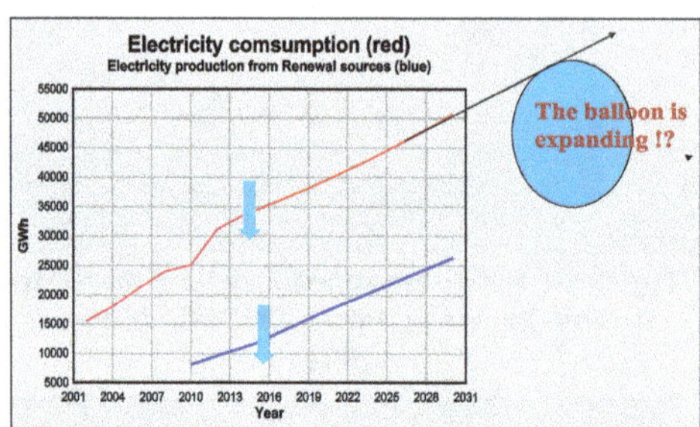

Fig. 9.3: Morocco Electricity Demands 2002-2030 (Top curve) and Electricity from Renewable 2015-2030 (Bottom curve). The arrow shows the year of the turning point by the introduction of the CPV Noor1 in the grid. P (107)

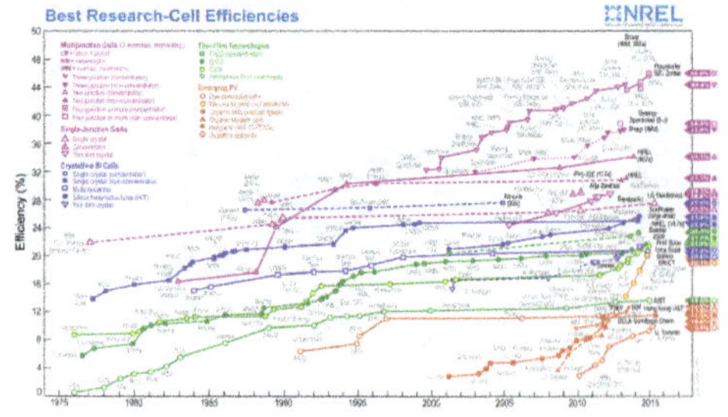

Fig. 10.1: NREL Ranges of different Modules Efficiencies. P (114)

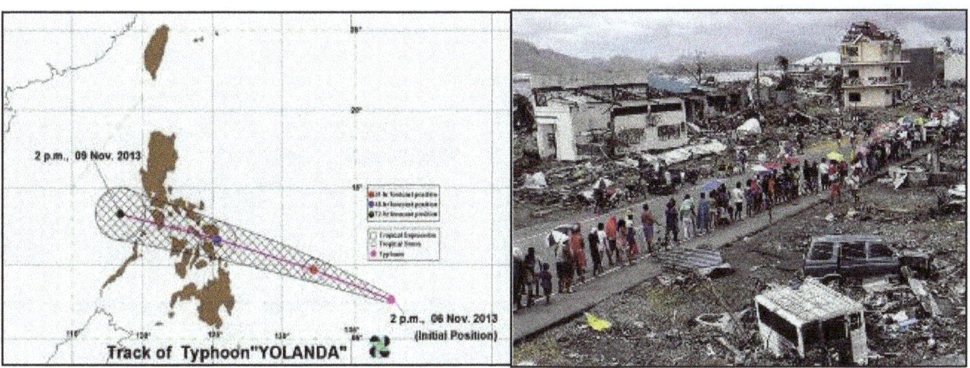

Fig. 11.1: *Right*–Track of Typhoon "Yolanda" that devastated the Visayan Regions of the Philippines. *Left*–People of Tacloban, Leyte fell in line for relief goods after the typhoon "Yolanda. P (121)

Fig. 11.2: Actual Footages of the Aftermath of Typhoon "Nina" Leaving at least One Month of No Electric Supply. The SES proved to be functional and beneficial to evacuees and the community in Bry. Yook, Buenavista, Marinduque. P (122)

Fig. 12.1: Provincial Map of South Africa. P (131)

Fig. 12.2: Witkop Solar Power Plant. P (132)

Fig. 12.3: Soutpan Solar Power Plant. P (132)

Fig. 12.4: Tom Burke Solar Park. P (133)

Fig. 12.5: Standalone PV. P (133)

Fig. 12.6: Vuwani Resource Science Centre Solar Power. P (134)

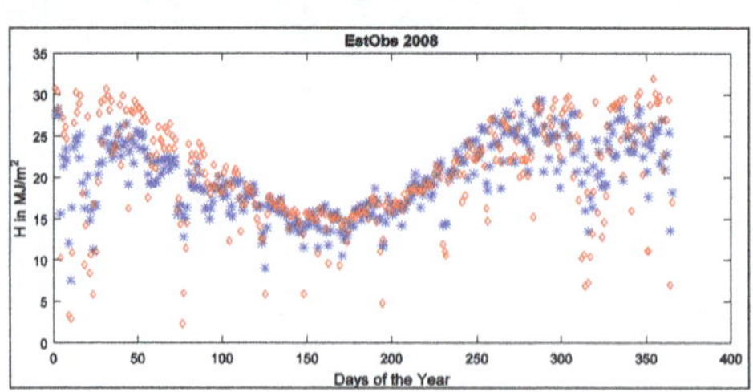

Fig. 12.9: Marble Hall – Comparison of the Observed and Estimated H. P (138)

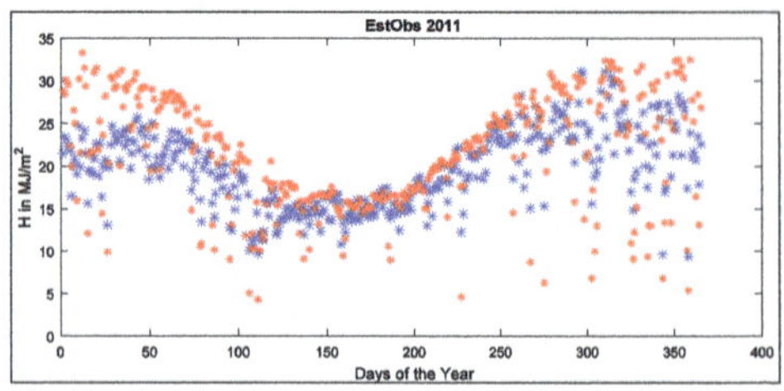

Fig. 12.10: Marble Hall – Comparison of the Observed and Estimated H. P (139)

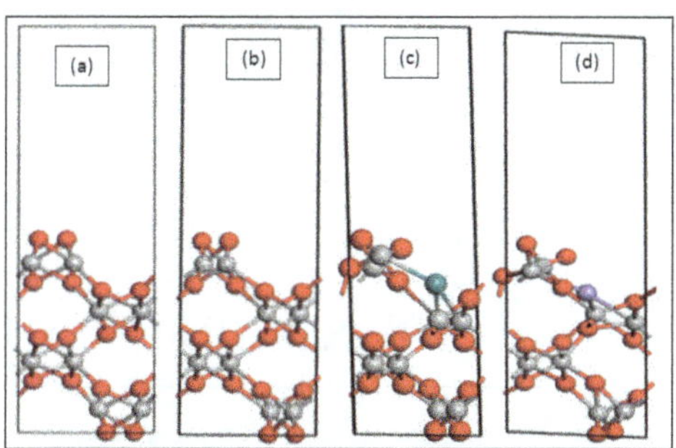

Fig. 12.11: Atomic Structures of (a) (100) Un-doped TiO_2 Brookite Surface before Optimization, (b) (100) Nu-doped TiO_2 Brookite Surface after Optimization, (c) (100) Ru-doped TiO_2 Brookite Surface After Optimization, (d) (100) Fe-doped TiO_2 Brookite Surface After Optimization. P (139)

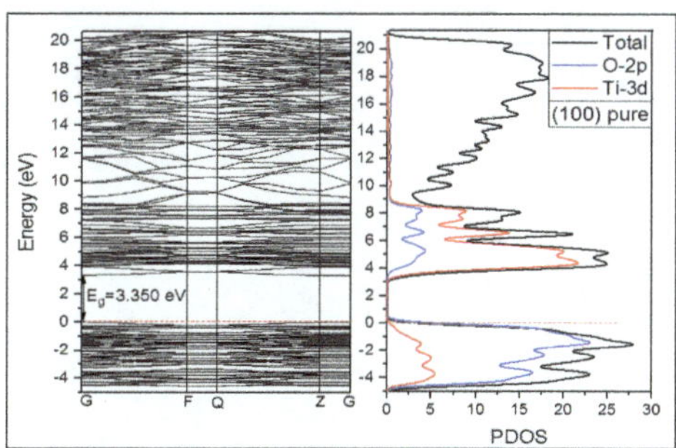

Fig. 12.12: Band Structure and State Density of Un-doped Brookite TiO_2 (100) Surface. P (141)

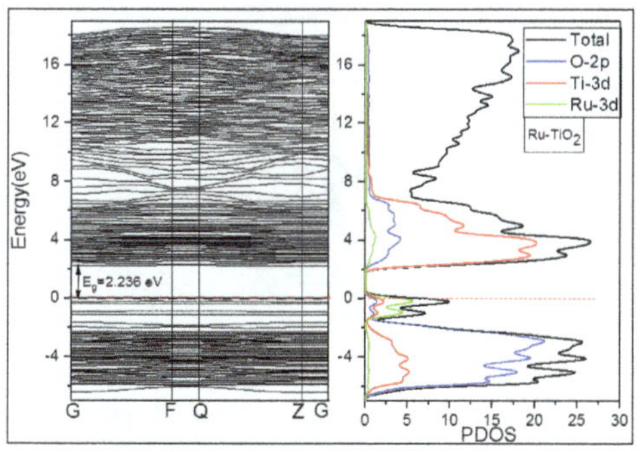

Fig. 12.13: Band Structure and State Density of Ru-doped Brookite TiO_2 (100) Surface. P (141)

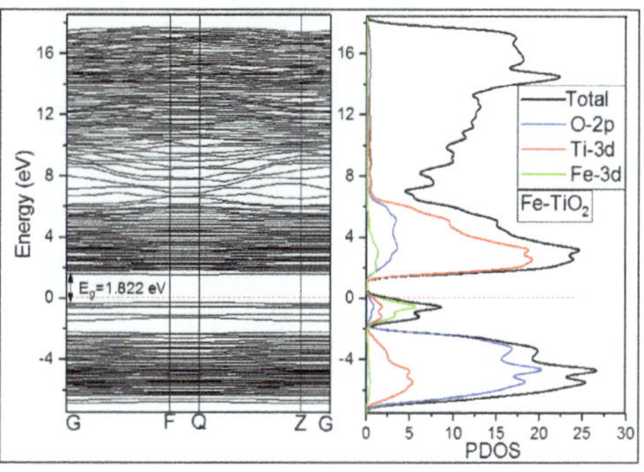

Fig. 12.14: Band Structure and State Density of Fe-doped Brookite TiO_2 (100) Surface. P (142)

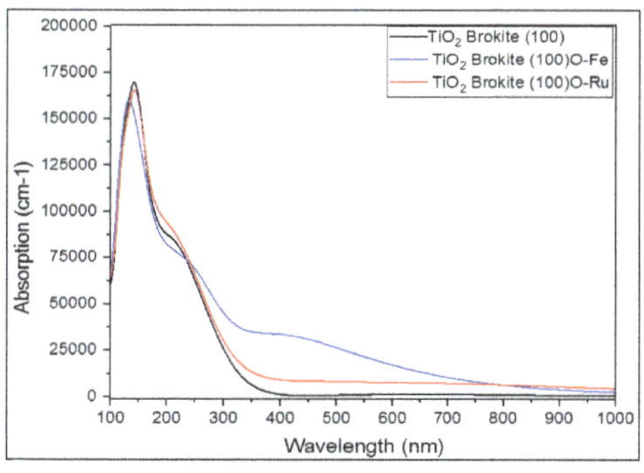

Fig. 12.15: Optical Absorption Curves of the Un-doped and Doped Brookite TiO_2 (100) Surface. P (143)

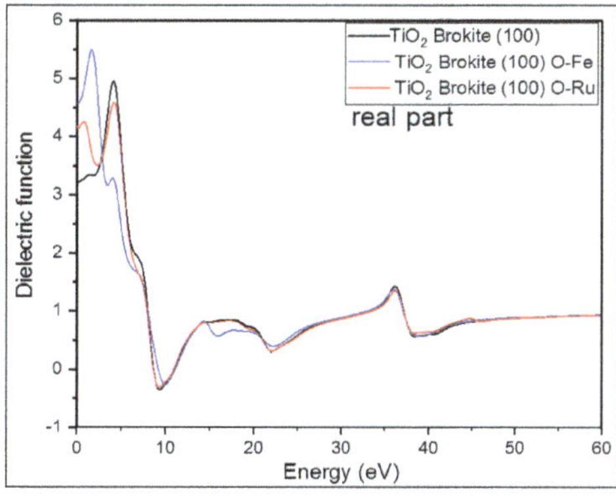

Fig. 12.16: Dielectric Function of the Un-doped and Doped Brookite TiO_2 (100) Surface. P (144)

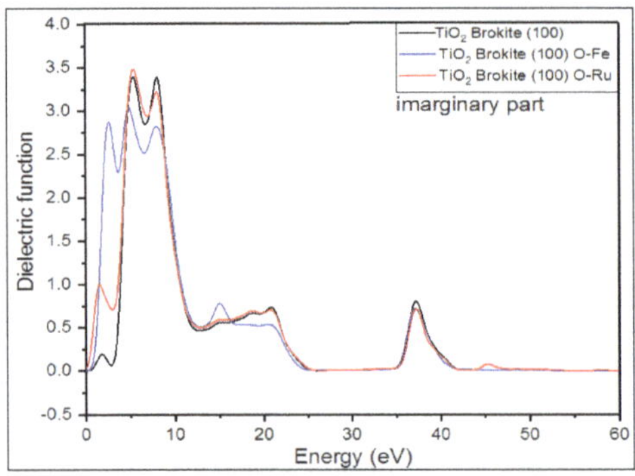

Fig. 12.17: Dielectric Function of the Un-doped and Doped Brookite TiO_2 (100) Surface. P (144)

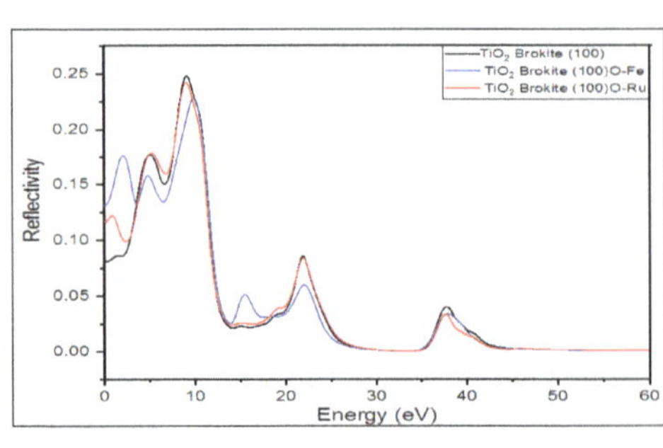

Fig. 12.18: Reflection Spectrum of the Un-doped and Doped Brookite TiO_2 (100) Surface. P (145)

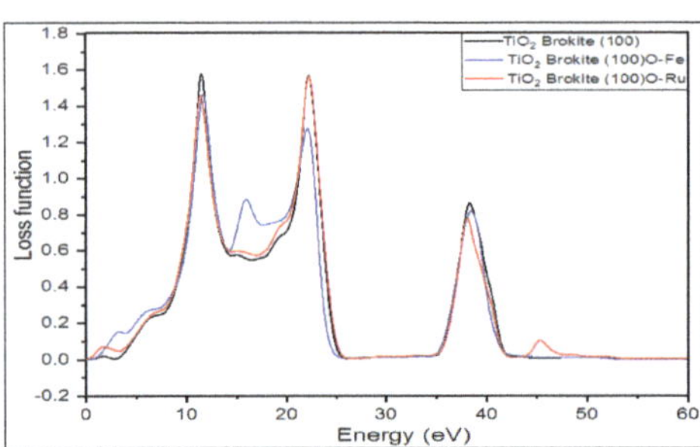

Fig. 12.19: Energy Loss Function of the Un-doped and Doped Brookite TiO_2 (100) Surface. P (145)

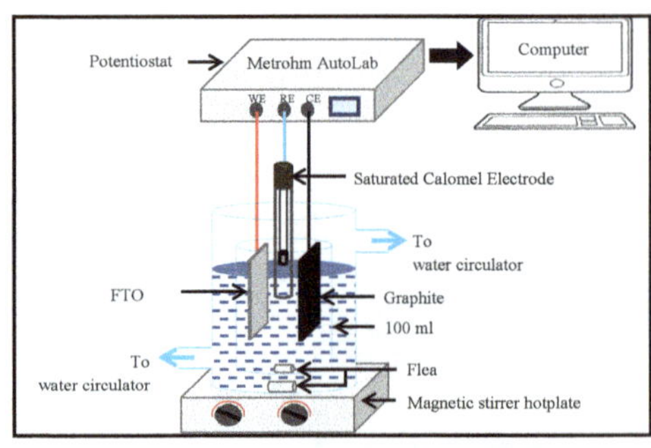

Fig. 13.1: Schematic Diagram of the Electrodeposition Setup. P (151)

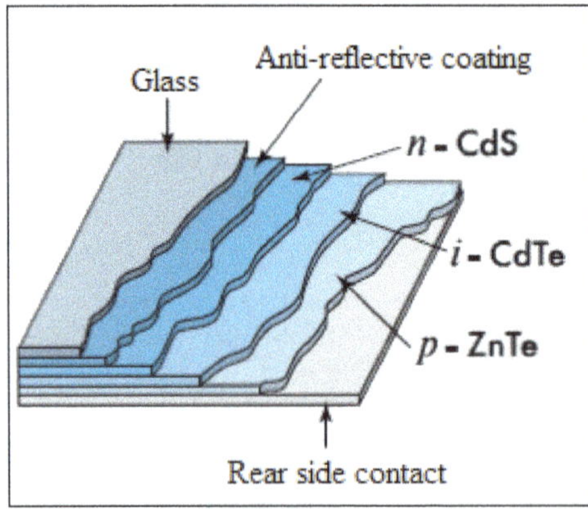

Fig. 14.3: Structure of a Solar Cell based on CdTe. P (160)

Fig. 15.2: Small PV System Installed at Mlalo Village in Muheza District. P (174)

Fig. 15.3: Powaga irrigation System in Rural Iringa District. P (174)

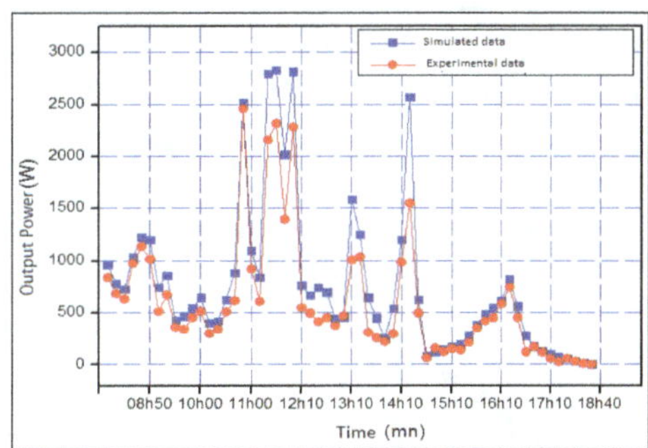

Fig. 16.4: Comparison between the Measured and Simulated Output Power Values for a Cloudy Day. P (186)

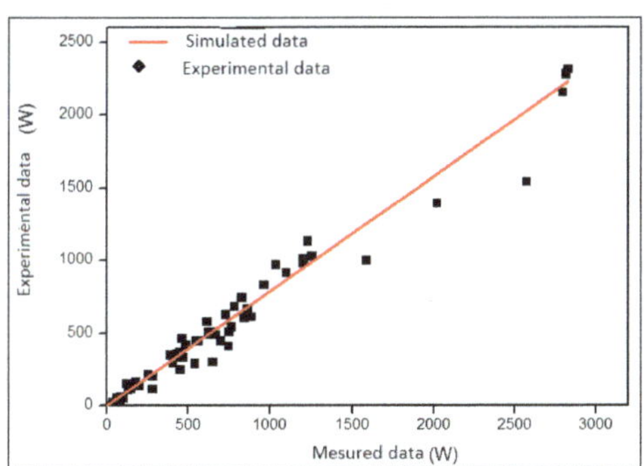

Fig. 16.5: Correlation between Measured and Simulated Power Values (Cloud conditions). P (187)

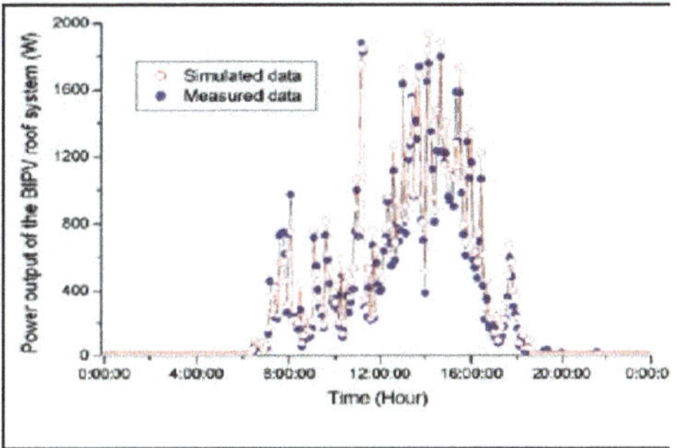

Fig. 16.6: Results Obtained by Whei Zhou *et al.* (Cloud Conditions). P (187)

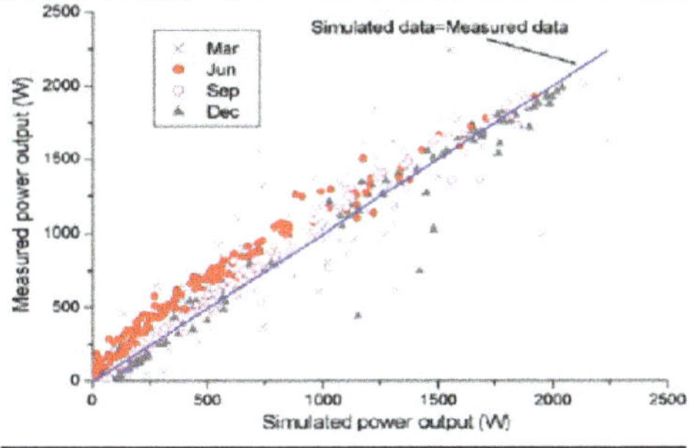

Fig. 16.6: Results Obtained by Whei Zhou *et al.* (Cloud Conditions). P (187)

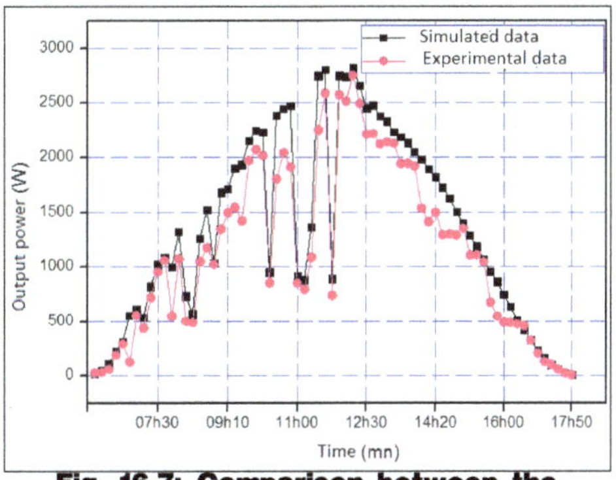

Fig. 16.7: Comparison between the Measured and Simulated Power Values (Clear Sky Conditions). P (188)

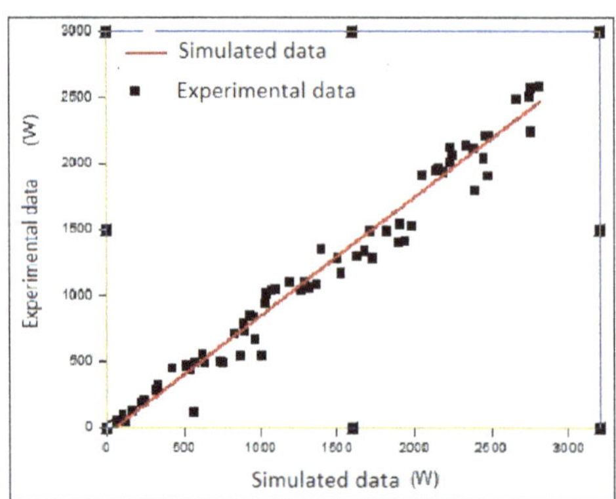

Fig. 16.8: Correlation between Measured and Simulated Power Values (Clear Sky Conditions in Togo). P (188)

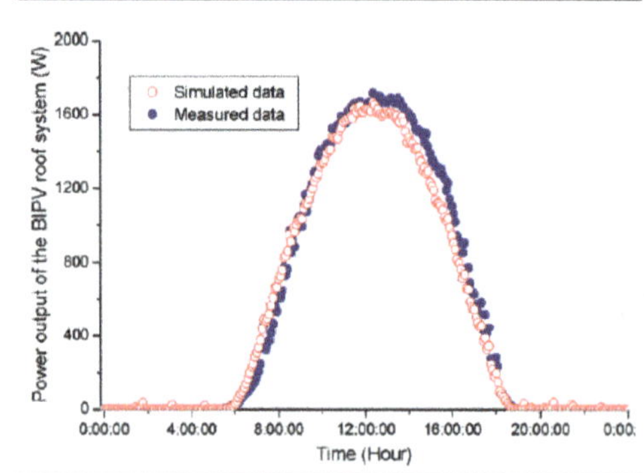

Fig. 16.9: Results Obtained by Whei Zhou *et al.* (Sunny Conditions). P (189)

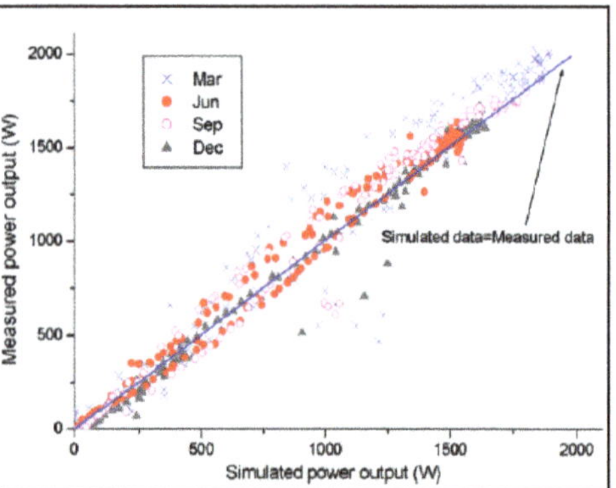

Fig. 16.9: Results Obtained by Whei Zhou *et al.* (Sunny Conditions). P (189)

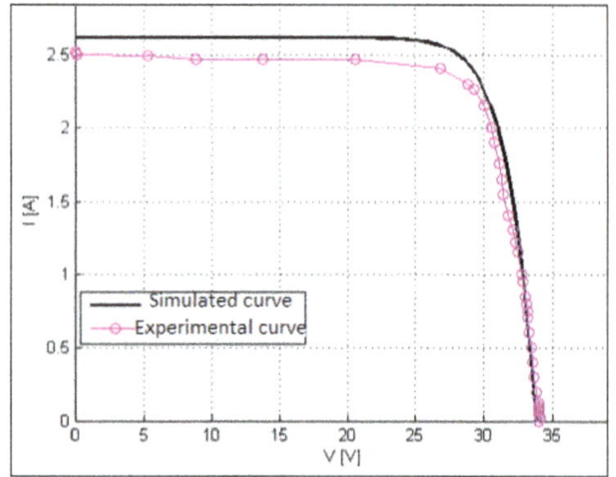

Fig. 16.11: Theoretical and Experimental I-V and P-V Curves for G=310W/mi; T=33°C. P (191)

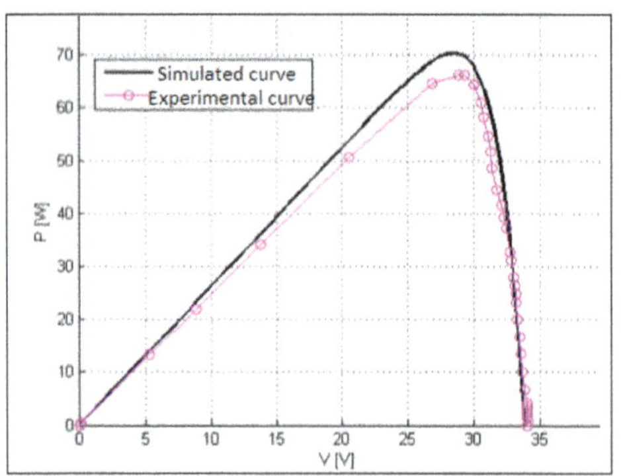

Fig. 16.11: Theoretical and Experimental I-V and P-V Curves for G=310W/mi; T=33°C. P (191)

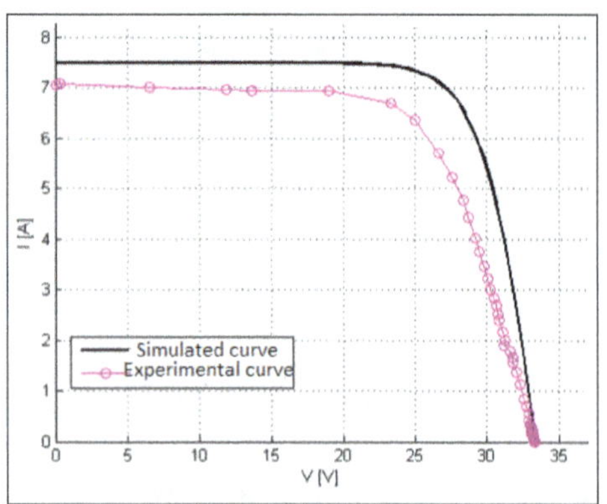

Fig. 16.12: Theoretical and Experimental I-V and P-V Curves for G=850W/mi; T=35°C. P (192)

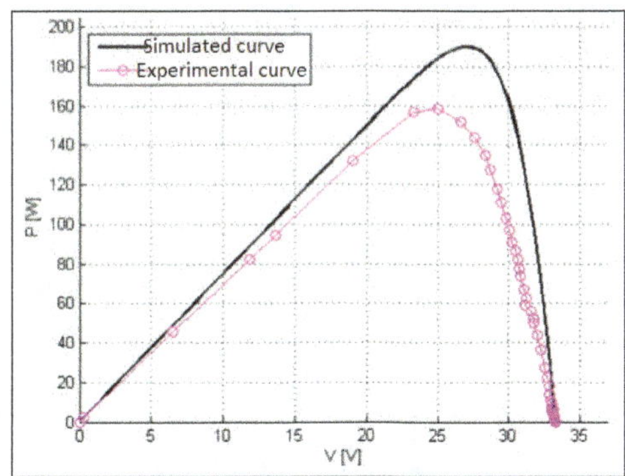

Fig. 16.12: Theoretical and Experimental I-V and P-V Curves for G=850W/mİ; T=35˚C. P (192)

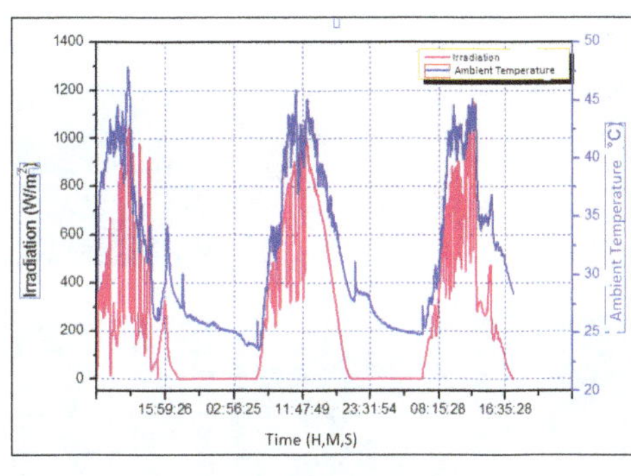

Fig. 16.13: Variation of Solar Dadiation and Ambient Temperature in the Vicinity of the Modules for Three Days at the Sūvagan Site. P (193)

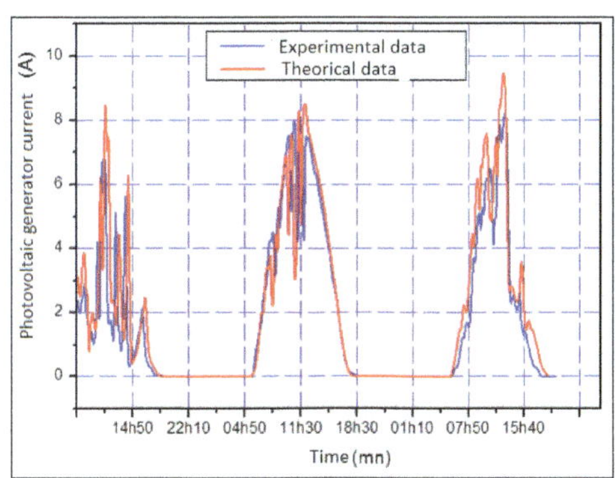

Fig. 16.14: Comparison of Measured and Simulated Photocurrent Values. P (194)

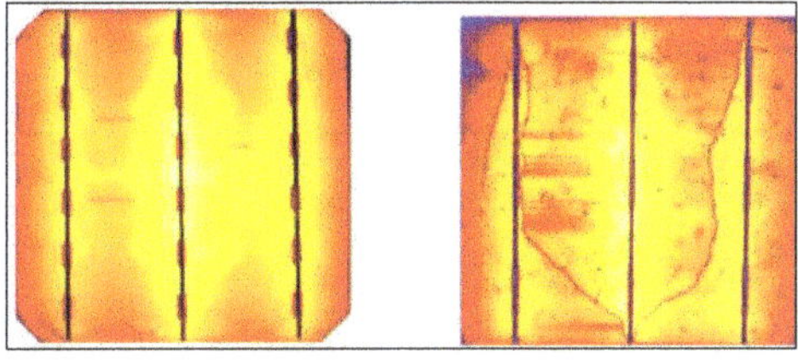

Fig. 17.1: Electroluminescence Images of a) Mono, b) Multi Crystalline Silicone PV Cells. P (201)

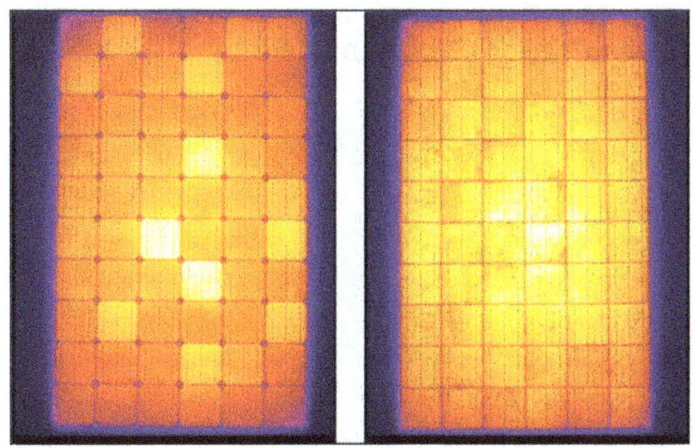

Fig. 17.2: Electroluminescence Images of a) Mono, b) Multi Crystalline Silicone PV Modules. P (202)

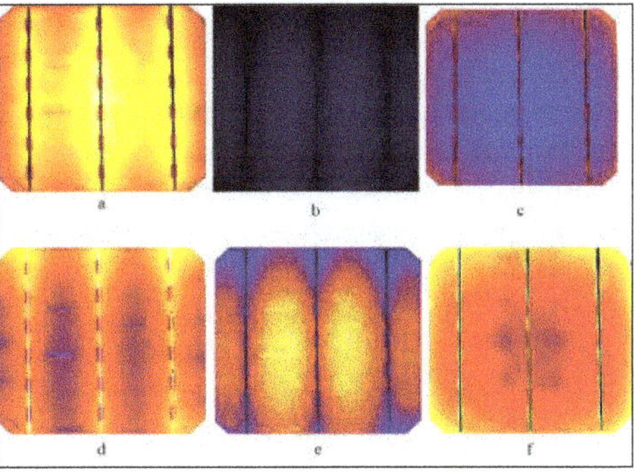

Fig. 17.3: Electroluminescence Images of Mono Crystalline Cell after 1th Iteration a) V_{OC}, b) R_S, c) J_0 Mono, after 20th Iteration d) V_{OC}, e) R_S, f) J_0. P (202)

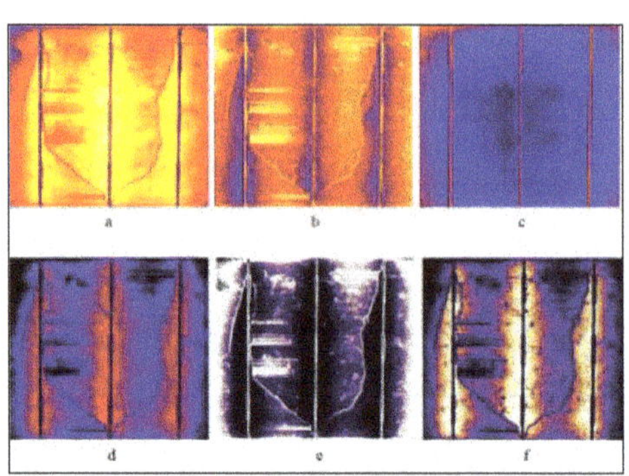

Fig. 17.4: Electroluminescence Images of Multi Crystalline Cell after 1th Iteration a) V_{OC}, b) R_s, c) J_0, after 20th Iteration d) V_{OC}, e) R_s, f) J_0. P (203)

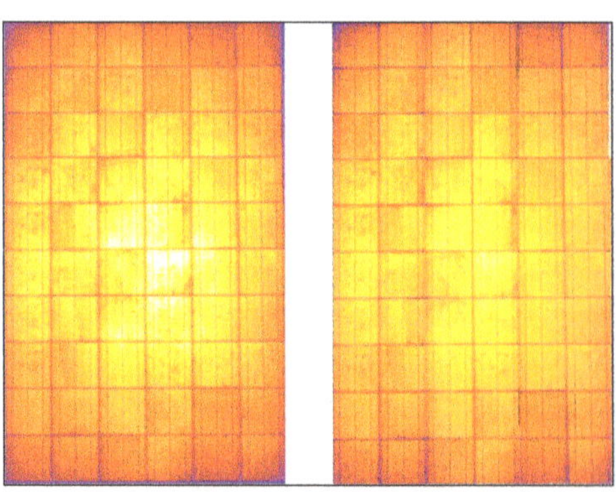

Fig. 17.5: Electroluminescence Images of a) One of the Multi Crystalline Module, b) Local Diode Voltages (Local Voc) Images Calculated Using Eqn. [3]. P (203)

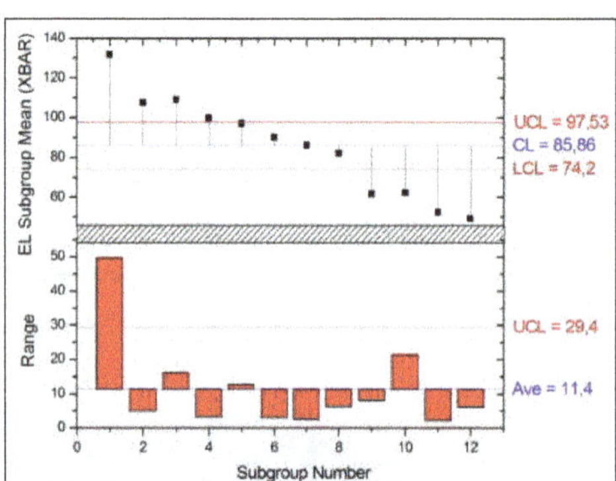

Fig. 17.6: Results of the SPCM Calculations a) EL Subgroup Distribution, b) Range of Distribution within the Group. P (204)

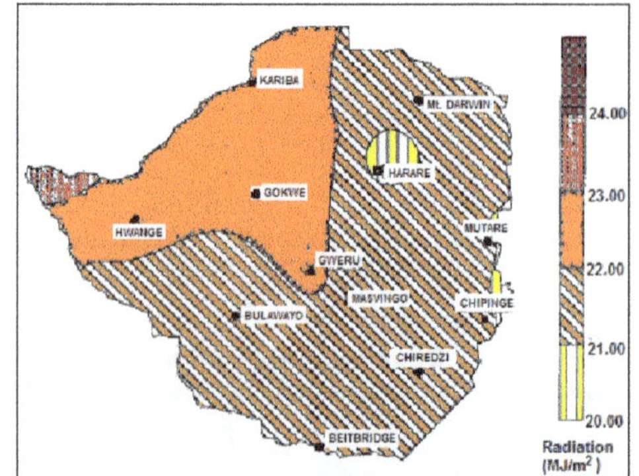

Fig. 18.1: Zimbabwe's Annual Mean Radiation (MJ/m²/Day) (*Source: ZERA*). P (208)

Fig. 19.3: Interface of the Software, Home Page. P (223)

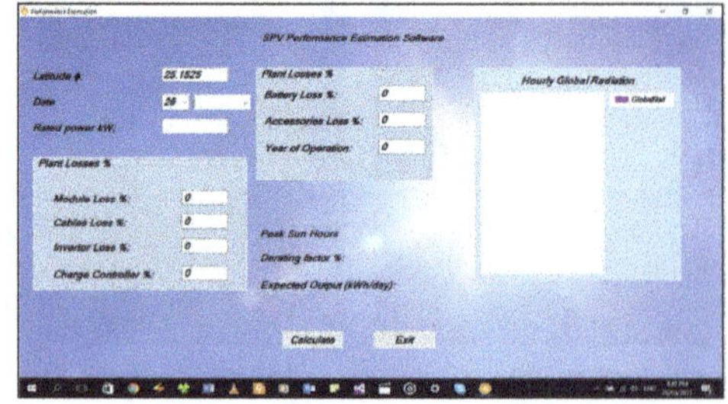

Fig. 19.4: Performance Tab without Inputs. P (224)

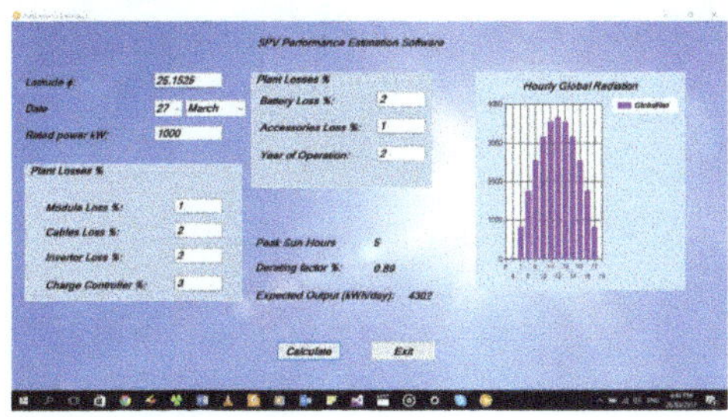

Fig. 19.5: Performance Tab with Inputs and Analysis Graph. P (224)

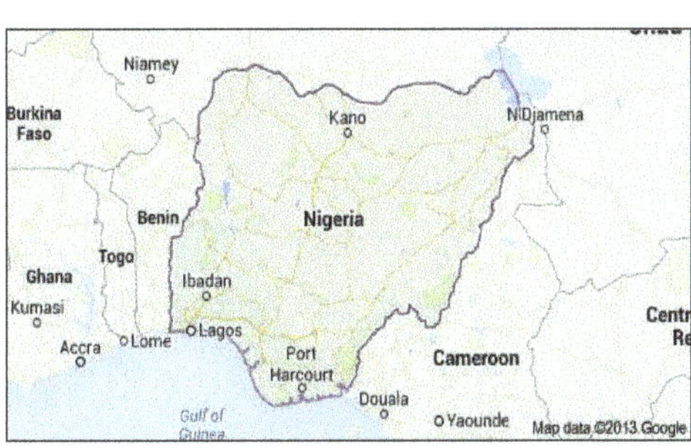

Fig. 20.1: Map of Nigeria Showing Boundaries. Google 2013 Map Data. P (228)

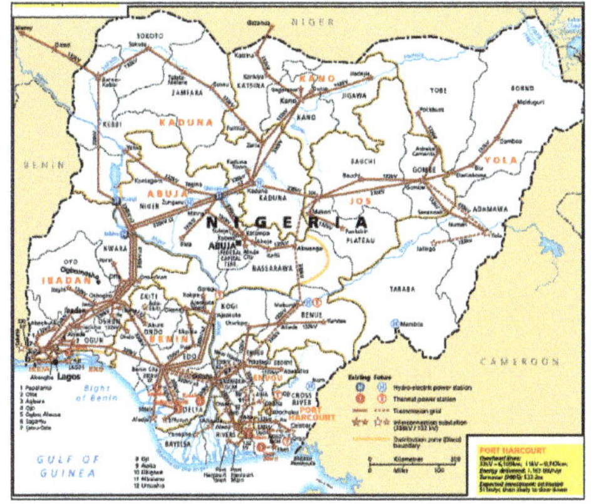

Fig. 20.2: Nigeria's Power Grid (*Source*: Google). P (230)

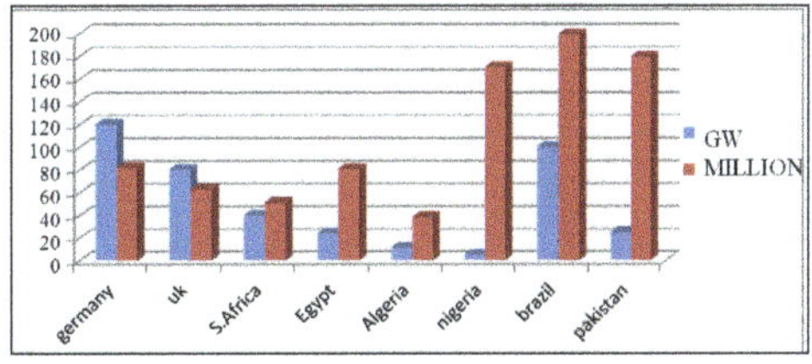

Fig. 20.3: Countries Population to Generation Capacity Ratios. P (230)

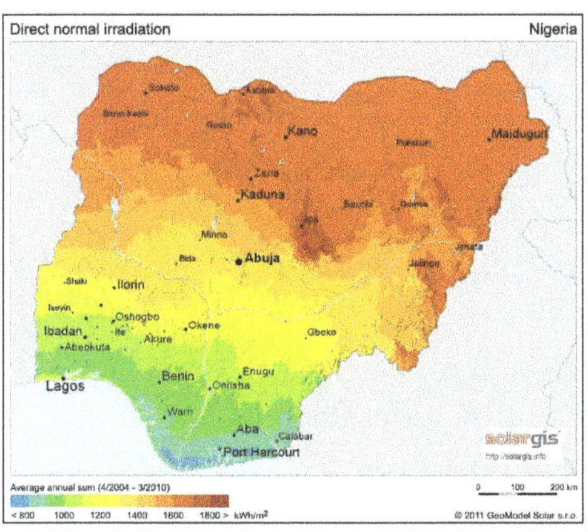

Fig. 20.4: Nigeria's Average Solar Radiation Map. P (234)

www.ingramcontent.com/pod-product-compliance
Ingram Content Group UK Ltd.
Pitfield, Milton Keynes, MK11 3LW, UK
UKHW021010290726
14059UKWH00001BA/57